India's Energy Security

This book explores the multifaceted aspects of India's energy security concerns. Bringing together a set of opinions and analysis from experts and policymakers, it sheds light on the context of India's energy insecurity and explores its various dimensions, its nature and extent. Contributors examine the role that trade, foreign and security policy should play in enhancing India's energy security. It is argued that the key challenge for India is to increase economic growth while at the same time keeping energy demands low. This is especially challenging with the transition from biomass to fossil fuels, the growth of motorized private transport, and rising incomes, aspirations and changing lifestyles. The book suggests that at this time there are strong arguments to lessen India's fossil fuel dependence and it argues for a need to engage with all the key sources of this dependence to implement a process of energy change.

India's Energy Security is a timely contribution given the national and international interest in the issue of energy security and the possibility that energy concerns could become the cause of serious international conflicts. It will be of interest to academics and policy makers working in the field of Asian Studies, Energy Policy, International Relations and Security Studies.

Ligia Noronha is a Senior Fellow and Director of the Resources and Global Security Division of The Energy and Resources Institute (TERI), New Delhi, India.

Anant Sudarshan is a PhD student in the Management Science and Engineering department at Stanford University, US and a visiting Research Associate at the Centre for Research on Energy Security at The Energy and Resources Institute (TERI), New Delhi, India.

Routledge Contemporary South Asia Series

1 **Pakistan**
Social and cultural transformations in a Muslim nation
Mohammad A. Qadeer

2 **Labor, Democratization and Development in India and Pakistan**
Christopher Candland

3 **China-India Relations**
Contemporary dynamics
Amardeep Athwal

4 **Madrasas in South Asia**
Teaching terror?
Jamal Malik

5 **Labor, Globalization and the State**
Workers, women and migrants confront neoliberalism
Edited by Debdas Banerjee and Michael Goldfield

6 **Indian Literature and Popular Cinema**
Recasting classics
Edited by Heidi R.M. Pauwels

7 **Islamist Militancy in Bangladesh**
A complex web
Ali Riaz

8 **Regionalism in South Asia**
Negotiating cooperation, institutional structures
Kishore C. Dash

9 **Federalism, Nationalism and Development**
India and the Punjab economy
Pritam Singh

10 **Human Development and Social Power**
Perspectives from South Asia
Ananya Mukherjee Reed

11 **The South Asian Diaspora**
Transnational networks and changing identities
Edited by Rajesh Rai and Peter Reeves

12 **Pakistan-Japan Relations**
Continuity and change in economic relations and security interests
Ahmad Rashid Malik

13 **Himalayan Frontiers of India**
Historical, geo-political and strategic perspectives
K. Warikoo

14 **India's Open-Economy Policy**
Globalism, rivalry, continuity
Jalal Alamgir

15 **The Separatist Conflict in Sri Lanka**
Terrorism, ethnicity, political economy
Asoka Bandarage

16 **India's Energy Security**
Edited by Ligia Noronha and Anant Sudarshan

India's Energy Security

Edited by Ligia Noronha and Anant Sudarshan

Routledge
Taylor & Francis Group
LONDON AND NEW YORK

First published 2009
by Routledge
2 Park Square, Milton Park, Abingdon, Oxon, OX14 4RN

Simultaneously published in the USA and Canada
by Routledge
270 Madison Avenue, New York, NY 10016

Routledge is an imprint of the Taylor & Francis Group, an informa business

Typeset in Times New Roman by Pindar NZ, Auckland, New Zealand
Printed and bound in Great Britain by TJI Digital, Padstow, Cornwall

British Library Cataloguing in Publication Data
A catalogue record for this book is available from the British Library

Library of Congress Cataloging-in-Publication Data
A catalog record for this book has been requested

ISBN 10: 0-415-46838-8 (hbk)
ISBN 10: 0-203-88436-1 (ebk)

ISBN 13: 978-0-415-46838-1 (hbk)
ISBN 13: 978-0-203-88436-2 (ebk)

Contents

Illustrations

Figures

Tables

Contributors

Talmiz Ahmad is currently Ambassador of India to the UAE. Mr Ahmad has been the Indian Ambassador to Saudi Arabia (2000–03) and Oman (2003–4), and Additional Secretary in the Ministry of Petroleum and Natural Gas. His book, *Reform in the Arab World – External Influences and Regional Debates*, was published in July 2005. He writes and lectures regularly on resurgent Islam, West Asian politics and energy security.

Ravi Kumar Batra is a Distinguished Fellow at The Energy and Resources Institute (TERI), New Delhi. He has 35 years' experience in the petroleum industry, beginning his career in Burmah Shell in 1957 and retiring as Marketing Director of Bharat Petroleum, a Fortune 500 company. He is currently on the Board of Directors of the Energy Institute, India, a chapter of the Energy Institute, London.

Koshy Cherail is Senior Consultant – Business Development with Conzerv Systems Pvt Ltd, India. Dr Cherail has more than 20 years of experience in programme implementation, policy analysis and consulting with various bilateral and multi-lateral agencies, including World Bank, USAID and GTZ. He has been associated with the various efforts to organize the ESCOs and EE business in India. He has represented Indian ESCO and EE businesses at various international fora.

Pradeep Kumar Dadhich is a Senior Fellow at TERI. He has a Ph.D (IIT Bombay) in Energy Systems Engineering and over 27 years of experience. He has also been a member of the International Energy Agency's Greenhouse Gas RD&D Executive Committee since 1998. He was one of the lead authors in the IPCC's special report on carbon dioxide capture and sequestration. He is currently the Head of the Centre for Policy and Regulatory Studies at TERI University, New Delhi.

Mitali Das Gupta is Associate Fellow with the AEI and also with the Centre for Research on Energy Security at TERI. She has completed her doctoral thesis from Jadavpur University, Calcutta. She was selected for the International Association for Energy Economics (IAEE) student award for the year 2001. Presently she coordinates the activities of AEI. At TERI she is working on energy security, climate change and trade and environmental issues.

Ravi B. Grover is a nuclear engineer and is concurrently working as Director, Homi Bhabha National Institute (HBNI), Director, Strategic Planning Group (SPG), Department of Atomic Energy (DAE), and Director, Knowledge Management Group, Bhabha Atomic Research Centre (BARC), DAE, India. Grover is a Fellow of the Indian National Academy of Engineering. He studied mechanical engineering at Delhi College of Engineering, nuclear engineering at BARC Training School and received a Ph.D from the Indian Institute of Science, Bangalore.

Eshita Gupta has been working as a Research Associate at The Energy and Resources Institute, TERI, India since February 2006. She is an economist with a postgraduate degree in economics from Delhi School of Economics. She has a number of publications in refereed journals and proceedings of international and national conferences. She has worked extensively in the areas of energy, with a major focus being work related to energy-security indicators, energy-market risks and energy-access issues.

Commander Gurpreet Singh Khurana was commissioned into the Indian Navy in 1988 and is a Missile Warfare specialist. Before joining IDSA as Research Fellow in 2003, he commissioned the Fast Attack Craft INS *Tarmugli* at Port Blair as its first Commanding Officer. His research interests include maritime security, strategic and defence issues. He has authored more than 25 journal papers and made presentations at many international maritime conferences in India and abroad. He is the author of *Maritime Forces in Pursuit of National Security: Policy Imperatives for India* (2008) published by IDSA and Shipra Publications. He is presently involved in a study of China's maritime strategy and naval modernization.

Sudha Mahalingam is currently Member, Petroleum and Natural Gas Regulatory Board of India. She is also a member of India's National Security Advisory Board, advising the Prime Minister on security-related issues. In 2007, Sudha Mahalingam was awarded the first K Subrahmanyam award for Excellence in Strategic Studies. Prior to her current assignments, she held the prestigious Nehru Fellowship at Nehru Memorial Museum and Library. An economist and lawyer by training, she specializes in reforms and energy security. Her focus areas are energy regulation, tariff setting and the geopolitics of energy security.

Deepti Mahajan is Research Associate at The Energy and Resources Institute, New Delhi. Here she is part of the Centre for Research on Energy Security. She has a graduate degree in International Relations from the University of Nottingham, UK, and an undergraduate degree in Journalism from Lady Shri Ram College, University of Delhi. She was earlier with Women in Security, Conflict Management and Peace (WISCOMP), New Delhi, where she worked on multi-track diplomacy, and issues at the interface of gender, security and peace building.

Ajay Mathur is Director General of the Bureau of Energy Efficiency, Government of India, and a Member of the Prime Minister's Council on Climate Change. As Director General of BEE, he coordinates national programmes and policies to enhance end-use energy efficiency in the country. Prior to joining the Bureau, he was President of Suzlon Energy Limited; and has headed the World Bank Climate Change Team in Washington, as well as the Energy Technology Division of TERI in New Delhi. Dr Mathur received his Ph.D from the University of Illinois. Dr Mathur is the co-author of three books, including the IPCC Special Report on Technology Transfer.

C. Raja Mohan is currently a Professor at the S Rajaratnam School of International Studies, Nanyang Technological University, Singapore. Raja Mohan was the Strategic Affairs Editor of the *Indian Express* in New Delhi, and the Diplomatic Editor and Washington Correspondent of *The Hindu*. He has a Masters degree in Nuclear Physics and a Ph.D in international relations. He was a member of India's National Security Advisory Board during 1998–2000 and 2004–06. His recent books include *Crossing the Rubicon: The Shaping of India's New Foreign Policy* (New York: Palgrave, 2004) and *Impossible Allies: Nuclear India, United States and the Global Order* (New Delhi: India Research Press, 2006).

Nitya Nanda is a Fellow with the Centre for Global Agreements, Legislations and Trade at The Energy and Resources Institute (TERI), New Delhi. He has published several articles and papers on issues of trade, investment and competition, in professional journals, edited volumes and newspapers. His recent work includes *Expanding Frontiers of Global Trade Rules: The Political Economy Dynamics of the International Trading System*, published in 2008 by Routledge, London.

Ligia Noronha is a Senior Fellow and Director of the Resources and Global Security Division of The Energy and Resources Institute (TERI), New Delhi, India. She has been Secretary of the Asian Energy Institute since 2005 and coordinator of REEEP South Asia. She has a Ph.D from the London School of Economics. She is a member of the GOI's Expert Committee on Climate Change, is on the External Review Committee for Shell International's Sustainability Reporting and on the Task Force (natural resources, environment, land and agriculture) of the Commission on Centre State Relations of the Government of India.

M. V. Ramana, a physicist by training, is currently Senior Fellow, Centre for Interdisciplinary Studies in Environment and Development (CISED) at the Institute for Social and Economic Change, Bangalore. He obtained his Ph.D from Boston University, USA. Over the last few years he has been studying the Indian nuclear energy programme, focusing on economics, safety, and environmental impacts. He is co-editor of *Prisoners of the Nuclear Dream* (New Delhi: Orient Longman, 2003).

Surya Sethi is Principal Advisor (Energy) to the Planning Commission, Government of India, and part of India's negotiating team on climate change at the UNFCCC. He has worked in some 30 countries worldwide in the field of infrastructure, capital markets and industrial enterprises across a variety of sectors. Most recently, as Chief Investment Officer at the International Finance

Corporation (IFC) Mr Sethi's Energy Portfolio included power, hydrocarbons, energy efficiency, renewables and climate change initiatives.

J. Y. Suchitra has a Masters in Economics from the University of Hyderabad and has been examining the economics of nuclear power for four years. She is at the Institute for Social and Economic Change, Bangalore.

Anant Sudarshan is a Ph.D student in the Management Science and Engineering department at Stanford University and a visiting Research Associate at the Centre for Research on Energy Security at The Energy and Resources Institute. He holds degrees in mechanical engineering from the Indian Institute of Technology, Delhi and Stanford University. His present research focuses on the economics of energy and environment policies.

Foreword

The issue of India's energy security is not a new subject – it has been discussed ever since the first oil price shock took place in 1973–74. The Indian economy at that time was in a precarious condition, and therefore the quadrupling of oil prices led to inflation in double digits and, at some stage, inflation rose to well over 20 percent. The political fallout of those developments proved to be terribly expensive, and one could even say that Ms Indira Gandhi's imposition of emergency in 1975 was in some ways the outcome of what happened in the energy sector over the 1973–74 period. This historical fact only establishes the reality that the concept of energy security is very much part of a larger socio-economic and political construct. Currently, the Indian economy is in a fairly strong position, and therefore the spike in oil prices that has taken place in recent months has not in any way destabilized the economic progress of the country, even though growth is expected to be lower this year than in the previous three.

Dependence on specific sources of energy is a function of access to different forms of energy, their prices and the stock of capital which may constrain substitution between one form and the other. For instance, the Indian Railways having gone in for large-scale use of diesel locomotives cannot in a short period of time switch over to, say, electricity that would be based on an indigenous source of energy – coal. Similarly, given the problems associated with inadequate and unreliable supply of electric power, the country has seen a proliferation of captive power generating units, which are heavily dependent on the use of diesel oil. The recent increase in automobiles and their extensive usage has also led to an increase in consumption of oil. With the capital stock existing and the inertia in the system, India's dependence on oil would continue to grow in the foreseeable future. This would certainly raise questions relating to the country's energy security in several respects. Firstly, upward fluctuations in oil prices could prove disruptive to steady economic progress. Also, if on the basis of geopolitical changes the physical supply of oil were to be disrupted for a short period of time, the reserves of oil available in the country at any point of time would not be enough to withstand disruptions in oil-dependent activities. Finally, with growing imports and high oil prices, increasing consumption could prove to be an unbearable burden for the Indian economy, which may result in lower economic growth and loss of welfare.

One particular dimension of energy security which does not receive the attention it deserves is the issue of energy security for the poor sections of society in India. This is a subject that is of relevance not only to India, but to several other developing countries as well. There are today about 2.5 billion people in the world who are still dependent on the use of biomass energy. At the same time there are 1.6 billion people who have no access to electricity. For people who do not have a proper energy supply for cooking, and, therefore, have to depend on poor quality biomass, energy security takes on a very different dimension than what would be experienced in the modern sectors of the economy.

As yet the global community has not found satisfactory answers to this large-scale problem. Some limited efforts have been made, such as supply of LPG at highly subsidized prices to the population in some of the mountain states of India. This has proved very effective in providing a clean and modern fuel for cooking in some of these regions and has also helped arrest deforestation, which took place earlier in these locations for supply of fuel-wood. In respect of lack of access to electricity, at least the basic needs of people in rural areas for lighting can be taken care of through the promotion of solar lanterns and solar torches, which is exactly what TERI is attempting to do in its campaign for Lighting a Billion Lives (LaBL). But the task at hand is gigantic and the resources available have made the energy security objective of poor populations across the world clearly insurmountable under current conditions. This dimension of energy security has been covered in the book, but clearly much greater work is justified if this challenge is to be met on a large scale in different parts of the world. In my view, this book would be of great value for policymakers and citizens of the world who now need to ponder the direction of developments in the supply and use of energy, where changes are due to take place for various reasons, but most importantly also as a response to the problem of climate change. If any sectors in any economic system are likely to undergo major changes because of climate change, it is those involved in the supply and use of energy. The drive to ensure secure supply of energy would be a major factor in this change.

R. K. Pachauri
Director-General, The Energy and Resources Institute (TERI)
Chairman, Intergovernmental Panel on Climate Change (IPCC)
September 2008

Preface

This volume originated in a conference on **India's Energy Security: Foreign, Trade, and Security Policy Contexts** organized by TERI, and the Konrad-Adenauer-Stiftung, New Delhi on 29–30 September 2006. Given the growing concerns on the energy front, it was felt that there was a need to understand the factors that could enhance India's energy insecurity, to explore various dimensions of energy security in relation to India, and to examine the role, if any, that trade, foreign policy and overall security measures policy should play in enhancing it. The Conference had both Indian and foreign participants from Germany, Japan and China. The contents of the book, however, go beyond that conference. We include some of the papers from that conference that focussed on India, revised and updated, and we add new material in order to ensure that the debate reflects a concern not just with the external dimensions and security of supply issues, but more frontally with issues of energy access, increasing energy consumption of certain groups, the provision of clean energy and available technologies, and the environmental sustainability of energy choices. The purpose of this book is therefore to bring together a set of opinions and analysis, from experts and policymakers, with a view to crystallizing the assessment of challenges and opportunities before the nation on a subject that will remain central for some time to come.

India's energy concerns are really in the midst of a perfect storm: growing import dependency and rising prices of the fuels that the country needs to import; the complex geopolitics around energy supply sources and the growing pressures of the global community to make emerging economies, including India, accept commitments to limit the emissions of greenhouse gases. In many ways, these forces should create a movement away from current energy paths. But such change cannot come easily, and India is locked into a path dependency that will require it to engage actively with long standing compulsions of domestic energy pricing, technology choices, institutions and perceptions. The key objective of this volume then is to highlight the internal and external dimensions of India's energy security scenario, the choices it is consciously making, and the room to manoeuvre that it possesses in which to address these concerns.

We are very grateful to the Konrad-Adenauer-Stiftung for its financial support for the 2006 conference, particularly to Jörg Wolff, its Resident Representative at New Delhi. We would also like to thank him and the Foundation for permission to use some of the material that has appeared in an earlier summary of the conference proceedings. Some of the material in this volume draws from research being done under the project 'Building an Energy Secure Future for India' supported by the Nand and Jeet Khemka Foundation and we would like to thank them for their financial support. We thank the Director General of TERI, R K Pachauri, for his guidance on energy security policy issues and support with this project. Dorothea Schaefter, Associate Editor, Asian Studies, Routledge, has been very supportive of this book project, and this volume would not have taken shape but for her encouragement. We would like to acknowledge her role in this. Our thanks to Kate Moriarty for copyediting the manuscript and to Saroj Nair who helped with formatting the manuscript and other formalities. And last, and most importantly, to the contributors who attended the conference in 2006 and have since revised their papers despite their many other very pressing commitments, and to the new authors for their contributions that have together shaped this volume.

Ligia Noronha
Anant Sudarshan

Part I

Understanding India's energy security concerns

1 Contextualizing India's energy security[1]

Anant Sudarshan and Ligia Noronha

In the last couple of decades, energy-related policy challenges have grown increasingly prominent in India. On the back of a number of driving factors, (population growth, economic growth and lifestyle changes, among others) energy consumption has risen and its concomitant concerns have grown steadily. The forces that have shaped the development of the Indian state and the Indian economy over the last two decades, and that have led to the recent years of high growth, have also changed the paradigms within which energy policy decisions are taken.

There are two overarching forces influencing the country's energy policy decisions and creating the challenges confronting policy makers. These are *energy and growth* concerns and *energy and poverty* concerns. The first set of problems includes the need to supply enough commercial energy to drive growth, tackle unsustainable consumption, and improve our ability to cope with high energy prices. The second set of forces arises from the pressures generated due to large energy inequitics, the need to manage the transition from traditional fuel sources to cleaner fuels, and the provision of lifeline energy required to eliminate poverty and provide a basic minimum standard of living to all citizens. This duality of challenges that is before large developing nations such as India contrasts sharply with the situation in both developed countries, and the least developed nations. India's large and growing middle class[2] has energy-related concerns that bear a much closer relationship to those in the developed nations than to those of the very poor. At the same time, the majority of the country's population remains poor and predominantly rural, with no access to clean and modern energy, and little ability to pay for such.

The spread of issues arising from energy and growth concerns spans both internal and external dimensions of national policymaking. Arising out of the need to spur economic growth is the task of increasing domestic production of different fuels, and of dealing with environmental challenges – both local and global. There is also the challenge of walking the geopolitical tightrope that an increasingly import-dependent India finds itself on. There are perceptions of high risk, fuelled by growing dependence on oil imports and high oil prices. The heavily politicized nature of the international oil trade also adds to risk. In recent years geopolitics has become a central concern in energy trade, characterized by a rising resource nationalism; the concerns of an increasingly volatile West Asia; the responses that

India's emerging energy ties create among countries of the West; and potential threats to energy infrastructure and transit routes. Energy issues are becoming the lens through which many foreign and trade policy initiatives are being viewed, and part of the language of new diplomacy. On the environmental side, concerns about fossil fuel use have been tightly linked with climate change and international pressures for a cleaner energy path. India's room to manoeuvre is thus increasingly being framed by these developments (Noronha 2007).

Yet alongside the concerns we have just mentioned (which in many ways are just as important for developed nations), there is a different set of challenges specifically posed by energy poverty. Energy access is a huge problem in rural India, where traditional biomass fuels still dominate the energy mix. The 55th round of the National Sample Survey of India (1999–2000) found that 86 per cent of rural households continued to use biomass in the form of dung cakes, firewood or wood chips for cooking. Even today only 5 per cent of rural households use LPG, and only about 43 per cent are electrified (TERI 2005). While the penetration of modern energy sources was greater in urban households, over 20 per cent continued to rely on firewood and wood chips, and fewer than half used LPG for cooking. The total contribution of traditional fuels to the primary energy mix remains very significant (Figure 1.1). The presence of energy inequities in India also leads to the question of how to manage the transition from biomass to electricity and cleaner fuels (probably fossil fuels). Such a transition, given India's vast rural population, poses a major challenge from the point of view of ensuring supplies.

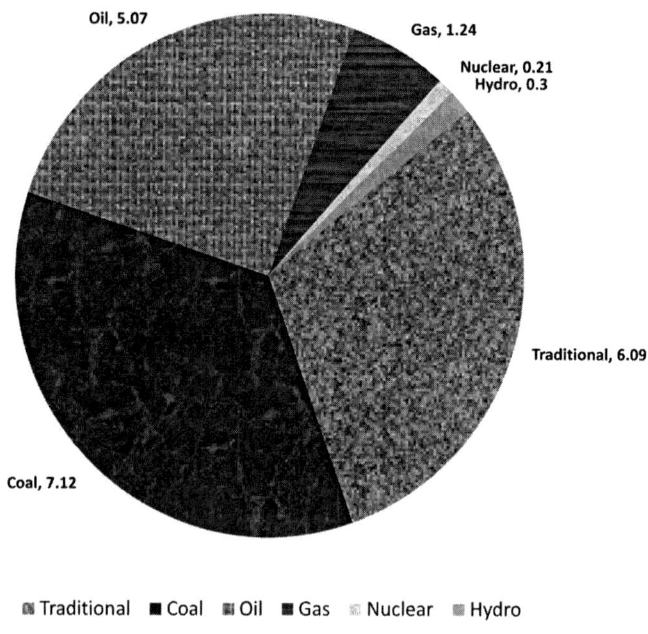

Figure 1.1 India's primary energy supply 2003–04 (in EJ).

(Source: Integrated Energy Policy Report (Planning Commission, 2006a.))

In this context, it is unsurprising that a number of energy-related concerns have arisen, for Indian people and policymakers. The term 'energy security' has become increasingly common, and defining an energy policy for India that takes us towards an 'energy secure' future has become an important goal today. Yet, for such an oft-used phrase, energy security has a somewhat ambiguous meaning. The sense in which the term is used has changed with time and so has its meaning in different parts of the world.

The concept of energy security

In April 1912, as Germany and Britain prepared to go to war, Winston Churchill – then the First Lord of the Admiralty – made the decision to switch the fuel used by the British Navy, from coal to oil. In doing so, he sought a crucial edge on the naval front, yet simultaneously committed the national security of Great Britain to oil supplies from Iran instead of coal from Wales. The company supplying oil to Britain – then called Anglo-Iranian oil and later becoming British Petroleum – became the object of investments by the British government and the rationale for an increased British military presence in the Persian Gulf.

The idea that energy policy, foreign policy and national security are all linked is therefore not new. Over the years however, conceptions of 'energy security' have evolved and grown rather more complicated. After the oil crisis of the 70s, as prices rose and a scarcity of affordable supplies became a real and pressing concern for countries in different parts of the world, ensuring energy supplies became a constant concern for all governments. Since then, energy security has been seen primarily as a problem of supplies and costs. In 1999, the United Nations defined energy security as requiring 'the continuous availability of energy in various forms and in sufficient quantities at reasonable prices'. Other organizations such as the Energy Information Administration of the US Department of Energy and the World Coal Institute use similar, supply centric definitions of energy security[3]. Along with an emphasis on supplies and costs, have come debates on issues such as the geopolitics of energy, the links between energy and foreign policy, maritime safety of energy supplies, play-offs between strategic energy investments and the development of free global energy markets, the nationalization of energy resources and the value of increased energy self sufficiency.

While all of these issues are certainly important, it has become increasingly clear in the last ten years that they provide only part of the picture. If the 'security' in 'energy security' is understood in the context of protection from energy-related threats, then the issues relevant to 'energy security' suddenly appear much broader. One way to understand this is to look at threats due to energy use as falling into the two categories mentioned above – those affecting economic growth and those that are unique to conditions of poverty. The traditional threats that form a part of energy security definitions – supply constraints, costs, import dependency, geopolitical tensions – all of these may be called energy and growth threats because they directly affect economic growth and the normal functioning of the economy. Other problems such as energy inequity, and the use of dirty, traditional biomass fuels with their associated health risks, could be treated as energy poverty threats.

Added to these challenges is the very major threat posed by global climate change. It is increasingly becoming clear that climate change is not only real and in large part caused by our dependence on fossil fuels (IPCC 2007a), but that its consequences will also pose a serious development challenge, particularly to the more vulnerable developing world (IPCC 2007b). It is still the case that energy security and climate change are normally regarded as distinct, sometimes competing, objectives. In recent years however the world has moved towards the realization that policies that focus on only one of those two aspects are probably doomed to being severely suboptimal. As such (and in line with the idea that energy security is best approached by first asking what the causes of insecurity are), it makes sense to speak of environmental threats as energy security threats. Not to do so is to miss the central fact that nations can have only one energy policy, and we might as well look to identify what is optimal across multiple attributes, rather than create separate policy goals (in this case 'reducing GHG emissions' and increasing 'energy security') that often conflict with each other.

It is in this context that India's 2006 Integrated Energy Policy report defines energy security as follows:

> The country is energy secure when we can supply lifeline energy to all our citizens as well as meet their effective demand for safe and convenient energy to satisfy various needs at affordable costs at all times with a prescribed confidence level considering shocks and disruptions that can be reasonably expected.
> (Planning Commission, 2006a)

This definition includes within it key aspects of energy security: those related to poverty and those related to growth. It may therefore be adopted as a reasonable definition for India with the understanding that the word 'costs' needs to be read as referring to not just monetary costs, but also externalities such as local and global environmental costs.

This understanding of energy security as being a broad concept is useful for researchers and policy analysts seeking to appreciate developing country priorities and energy policy decisions. For a country such as India for example, it is necessary to assess the marginal benefits involved in addressing any particular aspect of this broad definition and the opportunity costs involved in placing an alternative issue on the backburner. For example, addressing the issue of climate change must be played off against the need to tackle other threats – low energy access, poverty and disease (alleviating which requires development and increasing energy use), insufficient access to modern technology and consumer products (implying the need for industrialization) and so on. In some situations it might be felt that the marginal benefits from growth, energy access, cost reductions or mitigating local socio-environmental damages might dominate those obtained from addressing a climate change concern. In other cases, the reverse might be true. These kinds of tradeoffs should be the central concern of policymakers looking to create an energy secure policy for India. In addition, for energy policy analysts both within and outside the country, appreciating the existence of these tradeoffs and the fact that priorities will necessarily differ for the developed and developing world, is essential to evaluating the current situation and suggesting the measures most urgently needed.

Drivers of energy concerns in India

The last two decades have been, for India, a time of change and growth along many dimensions (Table 1.1). Many, if not all, of the changes have strongly affected our energy needs and certainly contributed to energy security concerns. Economic changes have probably received the most attention – both economic growth rates as well as structural shifts in the economy. Yet there have been other important drivers of change. These include demographic effects – population growth, urbanization, changing kinship structures and the demographic transition. There have also been strong increases in energy demand from the transport sector – linked to growing ownership of private vehicles and increasing travel distances. Finally, domestic energy use has seen a slow transition from traditional, biomass energy to commercial fuels. Managing this transition is likely going to be one of the country's greater challenges in the medium term.

There are also a number of external factors that contribute today to energy insecurity in India. These include high risk perceptions, fuelled by the extent of energy imports, the price of fuels in the world market, and geopolitical realities. Taken together, these factors influence perceptions of space in an international context and the urgency with which foreign dependence is sought to be reduced.

The factors highlighted in Table 1.1 have played and continue to play a large part in India's growing energy security concerns. Much can be said about each of these, but we will turn at this stage to an examination of the constraints that have held back the formulation of effective policy responses to these and other problems.

Our primary concern is that India is currently heavily 'fossil fuel' or 'carbon' dependent, seemingly committed to a 'traditional' path of development and energy use. Figure 1.1 illustrates this quite starkly. Virtually the entire primary energy mix is carbon based and, except for the significant share of energy that still comes from traditional biomass fuels, the remainder is almost entirely from coal, oil and natural gas.

Of greater concern than the country's current energy use patterns though, is its projected energy use. Not only are the fossil fuels likely to remain important in the commercial energy mix, but also, as traditional biomass use declines, the demand for electricity and fuels such as LPG will rise sharply in the residential sector. A number of energy models exist that seek to project future energy use, including those in the National Energy Map report (TERI 2006) and the Integrated Energy Policy report (Planning Commission, 2006a). While different models and projections have differed in their details, there seems to be uniform agreement among energy analysts that India's future in the next three to five decades is inextricably linked to high fossil fuel use. As an example, the Integrated Energy Policy report

Table 1.1 Factors aggravating the rise of energy security concerns in India

Factors increasing energy demand	*As indicated by*
Economic growth	Annual growth in real national income in India at a rate of 6 to 7%. Economic growth of over 6 % in the last decade and future targeted growth rates of 8 to 10% (Planning Commission, 2006b).
Changes in the nature of the economy	Economic reforms post 1991, accompanied by structural shifts away from agriculture and towards the services sector. Strong demand for infrastructure, housing, retail, media and entertainment services and IT.
Demographic factors	Annual rate of population increase of about 1.9% p.a. over the last two decades. Percentage of urban population is projected to rise from 25.5% in 1990 to over 40% by 2030 (United Nations, 2006). Changing kinship structures with smaller households.
Growing transport sector	Demand for petrol and diesel between 1980–81 and 2003–04 grew 7.4% and 5.7% p.a. respectively (Integrated Energy Policy, 2006). Rising share of road haulage in freight transport and rapid increases in personal vehicle ownership. Nearly 50% of oil demand comes from the transport sector (a share set to grow).
Incomplete energy transition	Biomass still the major cooking fuel in rural India, and less than 50% of rural households are currently electrified. In 2003–04 the domestic sector accounted for 25% of total electricity consumption, a share that is rising as fuel choices change.
Supply side pressures	India has 17% of the world's population, but only 0.8% of known oil and gas resources. Environmental concerns, inaccuracies in reserve estimates and coal quality concerns have led to constraints on domestic coal use, leading to rising import dependencies for all the fossil fuels.
External security concerns	Crude oil import dependency is projected to rise from 35% in 2001 to 78% by 2031 (TERI 2006). Imports of natural gas and coal are also expected to grow. India has been heavily dependent on West Asia for its oil and gas needs. More recently the focus of attention has shifted to sourcing from Africa. The growing oil import bill, as well as the projected increased dependency on imports for coal, oil and gas, is creating pressures for change in the way India engages energy-rich countries.

projects a maximum share of 4.5 per cent for renewable sources of energy, even if the country succeeds in developing 100,000 MW of renewable capacity over the next 25 years (which is estimated as an upper-bound on the feasible potential). Similarly, Figure 1.2 is a projection of India's commercial energy demand in a business-as-usual scenario and it illustrates both a sharp increase in expected energy demand, as well as a continuing dependence on fossil fuels.

Evidently there are serious concerns about both the environmental sustainability of this path and the degree of security it can provide in an energy stressed world. Therefore, for India to address energy security concerns, the country must

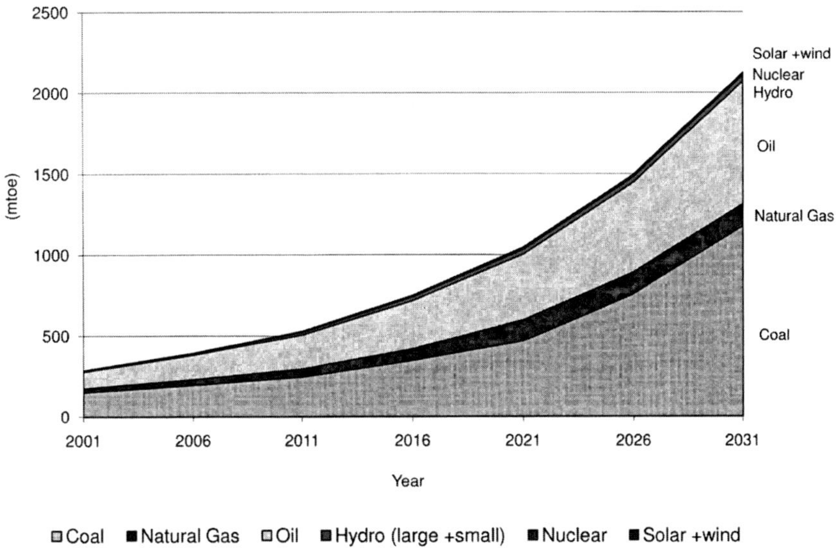

Figure 1.2 Rising commercial energy consumption in a BAU scenario.
(Source: National Energy Map for India: Technology Vision 2030 (TERI 2006.))

understand and tackle the underlying causes of this path dependence internally, while navigating the opportunities and risks that exist externally. To make our position on path dependence clear we quote here from North (2006):

> Path dependence is not 'inertia' rather it is the constraints on the choice set in the present that are derived from historical experiences in the past. Understanding the process of change entails confronting the nature of path dependence in order to determine the nature of the limits of change that it imposes in various settings.

Following North, we suggest that to understand what it would take to bring about change in India's energy sector, we need to focus on the sources of this path dependence.[4] We suggest that there are four key sources of path dependence in India's energy sector: *beliefs and perceptions*, *institutions and organizations*, *technology*, and *relative prices*.

Beliefs and perceptions are key to ordering mindsets and influencing choices. In the energy context, the beliefs that are important are those that relate to perceptions of resource availability, to what constitutes energy security, to the external 'symbols' of prosperity. If for example, both from the point of view of the state, as well as the consumer, energy security is perceived to be only a problem of *supplying* a minimal amount of energy, and of avoiding shocks and disruptions in that supply; then large parts of rural India are in fact reasonably secure as they are dependent on non-commercial biomass. If energy security is linked to the quality of the fuel and to related health implications of indoor air pollution, then this perception of security will change.

Whose perceptions matter and how these get translated into policy are a consequence of the institutional and political–economic structure of the society (North 2006). *Institutions* refer to formal rules, informal norms or constraints and enforcement characteristics. These are the rules of the game. How they are actually played out depends on *organizations*. These are groups of individuals bound by a common purpose to achieve objectives. They can range from the political, to the social, economic and educational. Organizations often depend for their survival on the non-alteration of rules and so exert pressures to avoid such changes. They may also play on perceptions and beliefs to strengthen the perpetuation of the system. The constraints to change from path dependence because of them are summed up well by North:

> Institutions that have accumulated give rise to organizations whose survival depends on the perpetuation of those instutions and which hence will devote resources to preventing any alteration that threatens their survival.
>
> (North 2006, 51–52)

In order to be energy secure we thus need to examine the institutions that we have in place and to ensure a more flexible institutional matrix that will adjust and adapt to evolving economic, political, technological and demographic changes as well as shocks to the system.

As far as *technology* is concerned, its creation, standardization, justification, deployment, modification, pricing, and promotion will determine its development, diffusion, and use (Dholakia *et al.* 1992). The availability of technology, learning capacities within a country, the quality of indigenous research and the funding of research and innovation efforts are all factors determining the viability of individual energy options. The context of energy technology choice making is key to understanding path dependence and performance, and the scope for change to alternative technology paths.

Relative prices of different fuels and incentives of various kinds are key determinants of individual choices. In the energy context, different subsidies and the non-inclusion of environmental externalities in pricing, have often led to choices being weighted heavily in favour of fossil fuel use.

We suggest that a movement away from the present pattern of energy use is constrained by a combination of these four sources of internal path dependency. To illustrate our argument we take up a few examples from the Indian energy situation – the continued dominance of coal in the power sector and of oil in the transport sector. The purpose of highlighting the forces that keep us fossil fuel centric, and some of the problems with our choices, is not necessarily to conclude that our current path is all wrong. Rather, it is to make the point that there has been an implicit acceptance of a state of affairs that needs to be questioned so that a more integrated policy may emerge. It is also to underline the fact that we need to think of alternatives to the current path for the long term.

Coal in the power sector

India's energy system has consistently been built around the heavy utilization of coal. Between 1996 and 2006, India's coal consumption rose from 154.4 million tonnes of oil equivalent (mtoe) to over 237.7 mtoe, recording an overall growth rate of about 54 per cent (BP Stats 2007). Today, thermal plants running on coal provide for 60 per cent of India's power generating capacity. The 2006 Integrated Energy Policy Report makes a number of recommendations, the first of which reads: 'Coal Shall Remain India's Primary Energy Source till 2031–32, Current shortages are a concern.'

In that context, it is interesting to understand not only how and why we have made this choice, but also what forms the basis of the underlying trend towards continuing in this vein. In light of concerns about global climate change and the possibility that India may have to take on some emission control obligations in the near future, a high dependence on coal may well prove something of a curse.

Technology

At the time of independence, India's total power generation capacity was 1,500 MW, mostly consisting of small hydro and high-grade lumpy–coal-fired thermal stations. The hydro: thermal generation mix was almost 50:50. Since India was, at the time, in a nation building phase, generation of electricity to nurture new industrial growth was the main task before the nation. For a poor country, the technology to use coal-fired thermal power plants was available easily, as were highly labour intensive processes for extracting coal through underground mining. Thus coal became a natural starter along with hydroelectric power. The pre-existing availability of a railway network for transport and a minimal amount of specialized requirements for transport was another initial advantage for the fuel. After nationalization, increased production requirements for coal could be met quickly by larger and larger opencast mines.

Unfortunately the evolution of technology in use in India has slowed down considerably. Even today, underground mining is inefficient and labour intensive and most thermal power plants use older, subcritical technology. The bulk of domestic coal comes from opencast mines and production increases are still driven by greater opencast mining (with little or no recovery of land afterwards) and over exploitation of existing mines. These technologies and processes are characterized by low present costs and quick increases in immediate production. They come at the cost of more sustainable methods (including efficient underground mining and land reclamation), and pose barriers to the technology learning and adaptation that need to take place before better mining technology can be adopted. India's experience with long-wall underground mining technology has been poor, largely due to a lack of skilled manpower, insufficient technical expertise, and the lack of research needed to adapt technologies from other countries to Indian conditions. In the long run, as we over-exploit shallow coal deposits, efficient and mechanized underground mining will become necessary. Unfortunately our continued reliance on short-term, sub-optimal technology today makes this transition rather difficult.

Relative prices

Until the 1980s, more coal was produced in labour intensive underground mines than in opencast mines. In the initial years, low wages helped keep coal production costs low, and in doing so helped make coal the mainstay of India's commercial energy generation. As the share of opencast mining increased, economies of scale again allowed coal costs to be competitive against other fuels. The presence of domestic reserves of coal remained a huge advantage – especially for a country short of foreign exchange. Coal can be extremely economical for use in pithead power plants, linked for short distances by rail. That said, because of the presence of heavy cross subsidies in rail transport, an archaic method of coal pricing based on useful heat value grades, and the fact that even today coal producers cannot really set prices free of government interference, it has been very difficult to assess the true costs of domestic coal compared with alternatives, including natural gas and imported coal. In effect, in the absence of proper markets, the status quo of coal dominance has proved difficult to change.

Beliefs and perceptions

Possibly the single most important reason underlying India's increasing reliance on coal over the last three decades, and its comparative neglect of other options, has been the idea that the indigenous reserves of coal in India are vast – plentiful enough to sustain the nation's need for hundreds of years. Even today, this impression has not entirely disappeared, both nationally and internationally. A normally authoritative source such as the BP Statistical Review of World Energy 2007 still listed India's proven reserves to production ratio as standing at a staggering 207 years. To a large extent the root of the ambiguity over the extent of India's mineable coal reserves stems from the fact that coal resources in India have been inventoried by the Geological Survey of India (GSI) on the basis of a geological classification system[5]. No comprehensive inventory has been carried out on the basis of the United Nations Framework Classification. As a consequence the exact amount of extractable coal in India is still difficult to ascertain. Various rules of thumb have recently been used to determine how much mineable coal India has, and the newest figures, such as those in the 2006 Integrated Energy Policy report, suggest that the figure could be as low as 45 years of coal at five per cent rates of production growth. Even this figure is likely to be an overestimate, since it includes sterilized coal, inaccessible coal, and some already exploited reserves. Whatever the real figure, it is clear that coal is far scarcer in India than many people imagined for a long time.

The harm this has done to India's energy policy over the years has been significant. The comfortable, opiating belief of possessing huge quantities of coal has contributed to a stagnation of energy policy initiatives and an insufficient investment in research and infrastructure that might have aided the use of alternatives such as natural gas or distributed renewables. Consequently today, not only is coal firmly entrenched as the primary fuel source in India, but alternatives are not really feasible immediately. This in turn means that new capacity increases must again be largely based on coal. This is important because power plants have a lifetime of about 30 years. Fuel choices today are decisions that have long-term implications. Thus choices made yesterday constrained us to a particular pattern for years; and decisions made today will continue to do the same into the future.

Institutional characteristics

Institutions and organizations play a major role in determining the course of government policy – both on paper and in practice. The coal industry was nationalized in India between 1971 and 1973 when first coking coal and then other coal mines were placed in the public sector. The reasons cited were that private operators were corrupt, engaging in slaughter mining, not willing to modernize, unable to increase coal production to meet national demand and were not meeting norms relating to working conditions. Unfortunately, today both coal production and the major consumption sector (power generation) are in the public sector. Coal production is a virtual monopoly, dominated by Coal India Limited (CIL) – a public sector behemoth with over 450,000 employees. As a result, the coal lobby in India is extremely powerful and any reforms that could force competition, reduce the importance of CIL, or change labour laws, have faced powerful opposition. A flourishing coal mafia and corruption have been difficult to remove in this situation and the status quo has consequently been maintained. Even the normal pressures from consumers that would exist in a market have been muted since both the largest consumer and the producer are state controlled. Thus reforms in other energy sectors have not spread to the same extent to the coal sector; competition and private participation remains miniscule; and best practices have not been adopted. Cleaning the Augean stables of the coal sector is therefore a prerequisite to actually achieving change on the ground. Unfortunately this is easier said than done when much of the change, and many of the recommendations needed, must come from the incumbent organizations.

Oil use in the transport sector

The demand for oil in India is almost entirely driven by the transport sector. Indeed, TERI projections (TERI 2006) show that, for a reference case scenario, the share of the transport sector in total petroleum product consumption increases from 36 per cent in 2001 to 64 per cent in 2031. It is clear therefore, that a business-as-usual approach to the transport sector has extremely worrying implications for the nation's future oil bill.

Technology

Unlike many other sectors consuming energy, the transport sector is characterized by very little flexibility when it comes to fuel choice. This is not a problem unique to India. The world over, high oil prices contribute first and foremost to increased costs of transportation. Highly fuel-efficient options such as hybrids are only now beginning to appear on the roads and are significantly more expensive than conventional IC engines. Similarly biofuels and fuel cell vehicles remain at a research and prototyping stage and are years from large-scale production. The technologies that drive our automobiles, ships and aircraft are well developed, mature and have proven both versatile and reliable. Unfortunately this has reduced investment in alternative engine technology that is less oil intensive, and correspondingly it is difficult and expensive to reduce our consumption of hydrocarbons for transport today.

India also suffers from the adverse consequences of having invested in urban transportation systems that are less than ideal. For example, the country's rail network remains largely non-electric, with only about 25 per cent of the route under electric traction[6]. Similarly infrastructure upgrading and technology leapfrogging has not occurred in road transportation – highways remain largely poor quality and insufficiently wide, and trucks and buses continue to use dated and inefficient designs. Thus the system remains *serviceable* without approaching anywhere near the desired quality and efficiency levels. As is common with many public goods, private investment is low and the state has not compensated – whether through the regulation of technology, the creation of incentives for the private sector or through a direct injection of resources.

Ultimately, a dependence on oil as a primary transport fuel is a global problem, having everything to do with the way transportation technologies and the supporting infrastructure has developed. However, it is necessary for India to play a lead role in moving away from this path, particularly since the country's demand for transport will only increase sharply. It is therefore in India's best interests to make sure that an atmosphere amenable to technological change is created.

Relative prices

The pricing of transport fuels and of modes of transportation has had a major role to play in determining the way things look in India today. For example, over the years, road transport has come to cater to more and more freight transport and passenger travel. This is in spite of the fact that rail transport provides significant economies of scale and can be far more easily electrified[7]. Now, while there are some advantages to road networks, such as greater flexibility and penetration, a large part of the reason for India's inability to exploit and expand the potential for railways as the preeminent mode of transportation has to do with a host of inefficiencies and biases in pricing.

The railways in India have consistently been burdened by social obligations and political agendas. This has led to measures such as cross-subsidization of passenger travel by much more lucrative freight transportation, resulting in a skewed pricing policy. The introduction of unviable lines driven by political whims and fancies, low employee efficiencies, involvement in a large number of non-core activities, and a limited investment in research and development (R&D) has compounded problems. In addition, rail transport covers all costs, including fixed infrastructure costs, in its pricing. On the other hand, road pricing does not reflect the full normative cost of ground infrastructure and its maintenance. This has adversely affected the market competitiveness of railways.

Finally, the presence of diesel and petrol subsidies and the reluctance of the government to allow fuel prices to be set by the market (even after the formal dismantling of the Administered Pricing Mechanism) have dampened the adjustments to demand that would otherwise have accompanied increases in oil prices, and have led to a 'dieselization' of the economy. The incentives to moving towards electrified rail transport, as opposed to diesel-based railways or road transport have been significantly reduced.

Beliefs and perceptions

State and consumer perceptions about public transport, their own entitlements and the best way to improve the system, play a large part in shaping the way the sector has grown. Urban transport provides an excellent example of this. To understand how, it is important first to recognize that the fundamental demand that needs to be addressed is for convenient and affordable mobility. While roads and private vehicles offer one way of obtaining this, there is a general reluctance to think of other ways in which the same ends can be achieved. For example, it is unfortunate that in India transport options such as bicycles, cycle rickshaws and walking are all seen as options only for the poor. Many cities in the world have encouraged choices such as these. Others have treated private vehicles as a luxury and stressed public transport (Singapore being one example). Encouraging alternatives that break away from the current paradigm requires a willingness to tackle the displeasure of the automobile lobby and of car owners.

It is possible for the state to implement policies that discourage and make difficult the ownership of cars; require a minimum number of passengers in some areas; make parking costly; and create automobile-free zones. Yet these options have been seen as less than desirable for a number of reasons. First, they place a certain responsibility on the state to provide substitutes – bicycle lanes, excellent public transport and better-designed cities. Second, they are seen as taking away from a fundamental right to enjoy *private* mobility and are thus restrictive. Changing these attitudes is important because it is only when the full social benefits of public transport are recognized that it will compete on a wide scale with options for private transport. The draft Urban Transport Policy says: 'In a developing economy, people have an urge to display their higher income status through the ownership of motor vehicles.' While there is an element of truth to this, it must also be said that consumer needs for private vehicle ownership must take second place to a more general need for mobility – at least where state policy is concerned. This distinction is unfortunately not made often enough and consequently we have an urban transport sector that excludes the poor and continues to grow in an extremely oil-intensive way.

Urban transport systems, and, to an extent, freight transport as well, can become classic examples of a tragedy of the commons. For instance, in Delhi, vehicle speeds went from 20–27 kilometres per hour in 1997 to 15 km per hour in 2002. In

Chennai, the average speed in 2002 was just 13 km per hour. In Kolkata the figure was 7 km per hour (Bidwai 2005). While it might seem to the typical middle-class consumer that his mobility needs are best served by a state that will widen roads, reduce vehicle costs and remove taxes on fuels, in reality such a policy risks spiralling towards a near complete breakdown of the system. Unless this is understood by the state, and communicated to the public, it is hard to see that improvement is likely in the near future.

Institutional issues

Most cities in India, with the exception of a few such as Chandigarh, have grown in an unplanned fashion. Rapid urbanization has increasingly led to the formation of satellite towns around the large metros and a heavy demand for transport over long distances from workplace to home. In such a situation, well planned and convenient public transport has immense potential. In the absence of such integrated transport systems, there will inevitably be a large increase in private ownership of vehicles.

Public transport in the country unfortunately suffers from poor quality and quantity. Most State Transport Undertakings responsible for public transport are not financially viable and suffer from losses. Their tariffs are low and have not kept pace with rising costs. Capital subsidies provided many years ago have now been withdrawn without any reduction in political pressure to keep fares low. Consequently, fleet maintenance and upgrading has suffered immensely. At the same time, the government has failed to provide an environment conducive to private sector participation. In the absence of efficient public transport, private vehicles have come to meet increasing mobility needs. Rising incomes, growing aspirations, liberalization of the auto market and easy financial support have led to a boom in growth of personal vehicles. In turn, this has made the automobile industry an extremely powerful lobby, influencing state policy.

The transport sector fulfils a consumer demand that is a prerequisite both for development and economic growth. In many ways, systems dominated by private transport and road transport are a default response to the lack of a well planned and integrated transport policy by the state. It is the failure on the part of the state to recognize how crucial is its role in the growth of oil consumption in transport that has led to the path India has taken. In all fairness, this is a phenomenon that has occurred in other parts of the world as well. Luckily for India, there is certainly enough time to push for a different vision of transport in the future, provided we begin to act today.

Before closing this discussion of oil use in the transport sector, it is important to highlight the fact that changes in technology, imposing efficiency standards, switching fuels and setting up better transport systems require strong political will. The changes also require the ability to make alternatives visible quickly so that people can see a coherent plan in place. In the long run, transport policy cannot and should not be determined by factors such as a court order to the government (as in the adoption by Delhi of CNG-based public transport). It is inevitable that lobbies and some of the stakeholders in the current system will oppose change that hurts them. The inability of the state to evolve consensus or to ignore these pressures is one reason why the transport sector continues to grow along the path of least resistance.

Conclusions

The term 'energy security' in the context of a country such as India refers to a complex concept and a number of issues. They include poverty, growth, and local and global environmental concerns. Over the course of this chapter, we have attempted to discuss how energy-security debates should be understood where developing countries are concerned, and have stressed the fact that many of the classical definitions of 'energy security' refer to a concept more suited to the developed west, than India.

In addition, we point out how there are a set of factors driving India's current energy concerns and that, as these forces play out over the next few decades, the country's energy challenges will only grow. It is in that light that we believe there is a real need for India to prepare itself to move off its current fossil fuel intensive energy path, which in general is not dissimilar to that followed by the developed world. Unfortunately, in the context of the world today, such a path is very likely unsustainable and increasingly insecure. With this in mind we look to ask why India has struggled to implement change, even where attempts have been made to do so. In order to carry out this analysis we draw upon new institutional economics to put together a framework through which we view two current areas of real concern – the continued dominance of coal as the primary fuel for power generation, and the slow pace of reform in the transportation sector. We show how both these issues can be seen through the lens of four constraints – *technology*, *relative prices*, *beliefs and perceptions* and *institutions and organizations*. Each of these challenges also forms part of our understanding of energy-security concerns and is directly linked to economic growth and the perception of development, the evolution of lifestyles and the extent to which India feels energy insecure as a nation.

Notes

1 This chapter is a revised version of the paper presented at the TERI-KAF Conference on 'India's Energy Security: foreign, trade and security policy contexts', 29–30 September, 2006.
2 Estimates of the size of the 'middle class' vary widely from about 50 to 250 million, depending on context and definitions.
3 The web sites of the EIA and the WCI as accessed on 10 March, 2007 underline the centrality of supply security to the idea of energy security. See http://www.eia.doe.gov/emeu/security and http://www.worldcoal.org/pages/content/index.asp?PageID=21
4 See North, 2006, chapters 5 and 6.
5 That is, resources are classified as proven, indicated or inferred based on geological considerations and not techno-economic feasibility of extraction. On the other hand, what matters from the point of view of energy policy is the amount of coal that is actually available for economic use.

6 By the end of the Ninth Plan, as per official Indian Railways statistics. See http://www.indianrailways.gov.in/railway/deptts/rly-elecfn/rly-elecfn.htm for more details.
7 Once electrified, the energy sources needed for transportation can be diversified and the variety of fuels that can be used for power generation come into play. Without electrification, the high energy density and portability of liquid fuels make them extremely difficult to replace.

References

Bidwai, P., 2005. 'Elitist recipe for more chaos.' *Frontline*, 22(15), viewed 15 May, 2008 <http://www.flonnet.com/fl2215/stories/20050729005711300.htm>.

BP Stats, 2007. *BP statistical review of world energy 2000.* British Petroleum.

Dholakia, N., Bakke, J. and Dholakia, R., 1992. 'Institutional patterns of information technology diffusion.' In Belk, Russell, W. and Dholakia, N. (eds). *Consumption and marketing: macro dimensions*. Boston: PWS-Kent.

IPCC, 2007a. 'Summary for Policymakers.' In Solomon, S., Qin, D., Manning, M., Chen, Z., Marquis, M., Averyt, K. B., Tignor. M., and Miller, H. L. (eds.) *Climate change 2007: the physical science basis*, Contribution of Working Group I to the Fourth Assessment Report of the Intergovernmental Panel on Climate Change, Cambridge: Cambridge University Press: Cambridge, pp. 2–18.

IPCC, 2007b. 'Summary for policymakers.' in *Climate change 2007: impacts, adaptation and vulnerability*, Contribution of Working Group II to the Fourth Assessment Report of the Intergovernmental Panel on Climate Change, p. 22, viewed 15 May, 2008 <http://www.ipcc.ch/SPM13apr07.pdf>.

MNES (Ministry of Non-Conventional Energy Sources) 1998, *Annual report 1997/98.* New Delhi: MNES.

NCAER, 1992, *Evaluation survey of household biogas plants set up during the Seventh Five Year Plan*, Vol. I. New Delhi: National Council for Applied Economic Research.

Noronha, M. Ligia, 2007. 'A India e o Contexto Energetic Internacional.' *Relacoes Internacionais*, Septembro., pp. 47–57.

North, D., 2006. *Understanding the process of economic change*, New Delhi: Academic Foundation, p. 6; pp. 51–52.

Parikh, J., Pandey, V., and Parikh, K., 2005. 'Lack of energy, water and sanitation and its impact on rural India.' In Parikh, Kirit S. and Radhakrishna, R. (eds.), *India development report 2004–05.* New Delhi: Oxford University Press.

Planning Commission, 2006a. *Integrated energy policy: report of the Expert Committee.* Government of India.

Planning Commission, 2006b, *Towards faster and more inclusive growth: an approach to the Eleventh Five Year Plan.* Government of India.

RBI, 2005, *Handbook of statistics on Indian economy*, Reserve Bank of India.

Sridharan, E., 2004. 'The growth and sectoral composition of India's middle class: its impact on the politics of economic liberalization.' *India Review*, 3(4), pp. 405–428.

TERI, 2003. *Rural energy matters: the Dhanawas experience.* The Energy and Resources Institute, New Delhi: TERI Press.

TERI, 2005. *TERI energy data directory and yearbook 2004–05.* New Delhi: TERI Press, p. 357.

TERI, 2006. *National energy map for India: technology vision 2030.* Prepared for the Office of the Principal Scientific Adviser, Government of India, The Energy and Resources Institute, New Delhi.

United Nations, 2006. *World urbanization prospects: the 2005 summary* (executive summary, fact sheets, data tables), Department of Economic and Social Affairs, United Nations, viewed 15 May, 2008 <http://www.un.org/esa/population/publications/WUP2005/2005WUPHighlights_Final_Report.pdf>.

2 India's energy challenges and choices

Surya Sethi[1]

In explaining India, I have often argued that no matter what one says about the country; exactly the opposite is true for a large section of her population. India is often presented as a major economy of the world, growing rapidly and poised to become the world's third largest economy by 2020. Yet India is also a country with the world's largest concentration of poor – over 830 million Indians live below the two-dollar-a-day level, with some 370 million of them living in abject poverty on less than a dollar a day. India's energy scenario also mirrors this reality. India, home to some 17 per cent of humanity, is the fifth largest consumer of fossil fuels in the world with a share of 3.7 per cent of the global commercial energy supplies, yet its per capita commercial energy consumption is only 20 per cent of the world average, 4 per cent that of the United States and about 28 per cent that of China. India faces many challenges in meeting the millennium development goals and raising its human development index. That said, water and energy are easily the two largest challenges in the country's path to emerging as a middle-income country and, importantly, there is a strong link between these two challenges. Addressing energy, this chapter begins by outlining India's energy challenge, provides a definition of an energy-secure India, details India's energy requirements and then outlines India's energy strategy.

The energy challenge

India's energy challenge is best summarized by the fact that some 600 million Indians live without electricity and over 700 million Indians still use traditional biomass as the primary fuel for the most basic human need, namely cooking. Finally, over 75 per cent of household energy demand results from the essential need for cooking energy. The backbreaking burden and drudgery of providing the energy for cooking through traditional biomass falls squarely on women and the girl child. Similarly, the health impact of burning this biomass is concentrated on these two groups, since they typically spend more time indoors.

Lack of a lifeline supply of safe and convenient energy is responsible for most of India's poor human development indices – either directly or indirectly. Illiteracy, gender inequality, disempowerment, high infant and maternal mortality rates, lack of access to clean drinking water, and poor health indicators can all be traced back to the lack of a lifeline level of safe and clean commercial energy for all Indian households.

India needs a consistent and *inclusive* eight per cent + GDP growth over the next twenty-five years to eradicate poverty and meet the millennium development goals. The emphasis on inclusive growth is essential if this aspiration is to be meaningful for all Indians. Raising access to modern commercial fuels is at the very heart of delivering such inclusive growth.

However, India must raise access to modern commercial energy in a sustainable manner, because even with a share of just four per cent of global GHG emissions, India is coming under increasing pressure to curb fossil fuel consumption. Thus India must decouple its economic growth from growth in energy consumption while ensuring universal access to lifeline levels of energy consumption. This is India's energy challenge.

Energy security in the Indian context

The World Energy Assessment (UNDP 2000) report defines energy security as the continuous availability of energy in varied forms in sufficient quantities at reasonable prices. This definition does not tell us how much is 'sufficient' or by what standard this 'sufficiency' should be determined. Further, it misses the concerns of sustainability.

In this author's view, India could be called energy secure only if the following criteria are met.

- The effective energy demand of all sectors for different needs is reliably met with safe, convenient and competitive energy in a sustainable manner. 'Effective demand' here means the demand at market-determined prices.
- Lifeline energy needs of all households are met with clean and safe commercial fuels where necessary, with transparent and targeted subsidies.
- All forms of available and emerging energy sources and energy technologies are adapted to achieve the first two goals above in a sustainable manner.

Clearly energy security in the Indian context, or the context of the developing world at large, is a complex issue that goes well beyond the typical understanding of energy security, which dwells upon assuring 'sufficient' supply at 'reasonable' prices. Such a traditional understanding of energy security was relevant only when the currently developed world enjoyed unconstrained growth by depleting the world's resource endowments and destroying our global commons. Today, however, the developing world cannot hope to get a scarce global resource at 'reasonable' prices and the definition of 'sufficient' varies with the level of economic development and resulting patterns of production, consumption and lifestyles. The increased use of commercial energy by those who still use more than a 'sufficient' requirement simply because they can afford to do so, threatens the very existence of those who never used it in the first place, or used it in insufficient quantities and at unreasonable prices. Thus, in the current context, energy security has strong overtones of equity and implies a moral responsibility towards reversing the historic impact on our global commons.

India's energy needs and threats

In 2005–06 India consumed about 513 million tonnes of oil equivalent (mtoe) (21.86 EJ) of primary energy, including some 367 mtoe (15.64 EJ) of commercial energy. The share of non-commercial energy, at about 28 per cent, was second only to India's most dominant energy resource, namely coal and lignite, which accounted for about 38 per cent of the mix. Oil and gas accounted for 24 per cent and 7 per cent respectively, with hydro and nuclear contributing 2 per cent and 1 per cent of the primary energy mix.

The 2006 Integrated Energy Policy report (see Planning Commission 2006) estimated that to sustain a growth of 8 per cent till 2031–32, India's total primary energy requirement would increase to about 1,536 mtoe in the most energy-efficient scenario[2]. Under the business-as-usual scenario India's total primary energy requirement would rise to 1,887 mtoe. This estimated requirement translates to an annual growth of 4.3 per cent in energy consumption over the 2003–04 level in the most energy-efficient scenario or an annual growth of 5.1 per cent in energy consumption over the 2003–04 level under the business-as-usual scenario. Should India succeed in meeting its economic growth targets with these estimated levels of growth in energy consumption, India's energy intensity and its emissions' intensity will have continued to fall.

The energy policy document estimated that the country's commercial energy requirement would increase more rapidly, at a rate of 5.2 per cent to 6.1 per cent over the 2003–04 level, to range from 1,351 mtoe in the most energy efficient scenario to 1,702 mtoe in the business-as-usual scenario. Thus India could lower its commercial energy requirement by over 20 per cent by following a more energy efficient growth path.

India's import dependence in 2005–06 was about 24 per cent of its total primary energy consumption in 2005–06 and about 33 per cent of its commercial energy consumption in the same year. By 2031–32 India's import dependence could rise to 28% of its total primary consumption, or 32% of its commercial energy consumption under the most energy-efficient scenario. Under the business-as-usual scenario, India's import dependence by 2031–32 would rise to 49% of the total primary energy consumption and 54% of the commercial energy consumption.

India's share of the world's supply of fossil fuels is projected to vary from as low as 3.7 per cent to between 7.6 per cent and 10.9 per cent by 2031–32, depending on whether one considers the most energy-efficient scenario or a business-as-usual scenario. Most importantly, India's incremental demand for commercial energy could account for 13 per cent of the world's incremental supply of commercial energy in the most energy-efficient scenario; rising to 21 per cent of the world's incremental supply of commercial energy in the business-as-usual scenario.

Such increases in India's share of the global commercial energy supply, especially on an incremental basis, will be extremely difficult to realize. Other developing countries that have registered high growth rates in the recent past, especially China, have already increased their share of the world's commercial energy supply and would also attempt to further raise their respective shares of any incremental commercial energy supplies. Given the overall supply constraints established by peaking oil and gas, future energy prices are likely to move northwards and remain there until major alternative sources and technologies come into play. For India, not following the most energy-efficient growth trajectory is simply not an option, if it is to deliver the inclusive growth that we talked about. And yet, even though energy needs have been estimated conservatively and even if India succeeds in delivering the energy efficient growth trajectory, realizing these levels of energy consumption is not going to be easy.

The domination of coal and lignite in India's energy mix will continue and is projected to rise to at least 41 per cent in the most energy-efficient scenario, or 54 per cent in the business-as-usual scenario by 2031–32. In fact, the share of fossil fuels in India's energy mix would rise from the 69 per cent level in 2005–6 to a range of 74–85 per cent, depending upon the country's degree of success in adhering to an energy-efficient growth trajectory. Thus, India's primary energy mix that had remained far less dependent on fossil fuels is likely to resemble the global primary energy mix more closely.

Most international forecasts project that the world's fossil fuel dependence is likely to rise from the 2001 level of 79.5 per cent to 84–85 per cent in 2031–32. It is hard to see a major technological breakthrough to reduce dependence on fossil fuels becoming widespread before 2031–32. Most breakthroughs are projected beyond 2050. The problem for India is getting to 2015, 2020, 2030 and 2050.

This author does not believe that current energy prices are too high or as disruptive as they were in the 70s and again in the 80s – a time when the global GDP was much lower than it is today. If we look at oil prices in constant 2003 dollars, the price of oil in the 12-year period from 1974–85 averaged USD54 per barrel. In 1980 the price was USD80 per barrel and it averaged USD72 per barrel during the period 1979–82. If one allows for the depreciation of the dollar and accounts for oil in a basket of currencies and recognizes that the global GDP today exceeds USD55 trillion, the current price of oil and gas does not seem that high. This is especially so when one also considers the supply constraints on conventional oil and gas. For these reasons, this author does not believe that crude oil is going to go below the level of USD85–100 per barrel in 2003 dollars over the next 10 to 15 years.

As pointed out above, the real issue is that India needs to grow her energy supplies and the share of world production of fossil fuels in a constrained market, wherein total demand is still rising. Thus availability or access to energy at an 'affordable' price is the first threat that India faces in meeting its energy needs.

As India, and, indeed, the rest of the world rediscovers coal, which is some 15 times more abundant than oil and gas, the second threat that India faces in meeting her energy needs is the pressure to reduce the country's incremental contribution to greenhouse gases. The developed world has chosen not to focus on the fact that their own greenhouse gas emissions are still rising in order to maintain unsustainable life styles, instead preferring to emphasize that India's share of the incremental GHGs is unsustainable. India will therefore have to go the extra mile to make the developed world understand that even under a business-as-usual scenario, India's per capita energy consumption in 2031–32 is expected to be below that of China's per capita energy consumption in 2005, about 16 per cent of the 2005 US per capita consumption and about 70 per cent of the 2005 average world per capita consumption. This underscores India's responsible use of energy even under the business-as-usual scenario. This is the outcome of India's frugal lifestyles and of energy prices that are the highest in the world in terms of purchasing power parity. If the multilateral community is committed to the eradication of poverty and to delivering the millennium development goals then it must ensure that India gets the modest levels of commercial energy projected to meet her needs in 2031–32.

The third major threat to India's energy security is the lack of a domestic energy infrastructure. As India grows its primary energy supply to the projected levels, it will need to also grow its energy-related infrastructure, such as power generation, transmission and distribution capacity, coal, oil and gas imports, transport and distribution capacity, associated rail, road and port facilities, as well as associated manufacturing, engineering and construction capacities. These supporting capacities need to grow three to eightfold in the next 25 years if energy supply and energy consumption targets are to be met.

Having identified the three key threats to India's energy security, let me address some other threats that are traditionally mentioned but to my mind are of a lower order of relevance to us. If shipping lanes for world energy routes get disrupted, or if political events cause disruptions in sources of supply, many other, more dependent, countries will hurt more than India and are also better placed to resolve such conflicts. Larger concerns at home limit India's ability to either cause or solve major geopolitical events. India's strategy can only be to react to, and merely provide for, such eventualities in its energy strategies.

Equity oil/gas in Sudan, Nigeria, Kazakhstan or Ecuador is not likely to come to India's rescue if the world truly faces an energy crunch or if political events disrupt supplies. Equity coal in Australia or Indonesia, which this author has been propagating since 2002, might be far more secure from an energy-security perspective. Suffice it to say that when Exxon Mobil makes overseas investments in oil or gas, it is not doing so to enhance the energy security of the United States. These are commercial investments and should be seen only as such. India's oil majors should make such investments primarily in the commercial interest of their shareholders. Such investments promote energy security only to the extent that they help diversify supply sources and promote energy diplomacy. Transnational pipelines add a larger degree of energy security through diversification of energy mix, energy supply infrastructure and supply sources. Pipelines cannot be diverted easily and the source country typically has a vested commercial interest through its ownership interest in the pipeline, and, potentially, in the downstream gas-based facilities.

Energy alternatives and energy choices

Given the fact that no technological breakthrough is being forecast in developing new energy sources by 2031–32, and given that India's growing energy demand is likely to account for a significant share of any additions to the world's fossil fuel supply; it is not a matter of choosing which source of energy India should tap. Thus India would need to pursue all possible energy sources and all available and emerging energy technologies that can make the use of fossil fuels more benign in the interim, while seeking technologies that provide potential non-fossil options for the future. Raising supply and demand side efficiency, expanding the use of renewable energy, switching out of coal and oil to gas and, where feasible, clean coal technologies, and greater dependence on nuclear power, are the major options that are currently engaging experts as practical alternatives for delivering sustainable development over the next few decades. Within that context, we look at a few key options for India to focus on in the next few decades.

Energy efficiency

The foremost 'supply' option in India's energy strategy remains energy efficiency and energy conservation. This acts as a virtual source, as it has the potential of reducing India's energy need for the same level of economic wellbeing by at least 20 per cent. India's energy intensity (ratio of total primary energy consumption to GDP in purchasing power parity (PPP) terms) in 2005 was 0.15. This ratio stood at 0.22 for China, 0.21 for the US, and 0.47 for Russia in 2005. India consumes far less energy per unit of GDP than most other countries. India's current energy intensity of GDP is less than half of what it was in the 70s and, more importantly, the point of inflexion occurred at a GDP level that was well below that at which the developed world started lowering its energy intensity. There is no other country in the world that has delivered an 8 per cent + annual GDP growth over the five-year period ending 2006 with just a 3.7 per cent growth in annual energy consumption. With 3.5 times the US population and three times the population of EU20, India has, since 2002, delivered more than twice their growth while consuming lower amounts of fossil fuels on an incremental basis in absolute terms – to reiterate, in absolute terms and not in per capita terms. China has grown faster than India but has also consumed over nine times the fossil fuels compared to EU20, over 10 times the fossil fuels compared to the US and over 11 times the fossil fuels compared to India on an incremental basis since 2002 in absolute terms. In fact, China's incremental fossil fuel consumption since 2002 is about 130 per cent India's total fossil fuel consumption.

This is not to say that India cannot do better. Although India's energy intensity shows up at the same level as Japan and Brazil and is better than that of Germany, India has paid a very heavy price for this achievement by denying access to modern commercial fuels to over half its population. This is indeed a heavy price to pay and is reflected in India's poor human development index. While Indian industry has made massive strides in achieving global energy efficiency standards in a number of energy intensive sectors, India can and should improve energy efficiency in a number of sectors both on the supply side and the demand side with particular emphasis on industry, buildings, and transport. India must also consider the fact that it should attempt to catch up with the United Kingdom, with an energy intensity of 0.14, and Denmark, at the top of the list, with an energy intensity of only 0.12. India has room to improve her overall energy efficiency performance by at least 20 per cent to match the best levels of efficient energy use based on currently available technologies.

Energy efficiency and energy conservation are together the most potent tools to ensure India's energy security as they can, virtually, meet some 20 per cent of India's potential energy needs.

Clean coal technologies

Coal, both domestic and imported, emerges as a key energy option for India. India must take the lead in clean coal technologies, especially in-situ coal gasification that can double the life of domestic coal from the estimated 40–45 years based on conventional mining. Tapping coal-bed methane, a well-established technology, can more than double India's gas reserves and supplies in the near term. Integrated gasification combined cycle (IGCC) technology using Indian coal has not been successful, but as India is going to become increasingly dependent on imported thermal coal, IGCC based on imported coal is a major option for raising conversion efficiencies and reducing the carbon intensity of power generation. Similarly circulating fluidized-bed combustion (CFBC) technology for the use of low quality coal and lignite can also enhance India's energy options and choices.

Expand hydrocarbon supply options

India is not known to have significant oil and gas resources, although some forecasts show significant endowment with gas hydrates. Given that oil and gas are likely to remain India's second largest energy source, India must fully exploit all onshore deposits, especially in the North East, seek deep-sea oil and gas under open acreage and pursue alternate routes to hydrocarbons based on tar sands, ore emulsion, methanol from marginal gas fields and gas hydrates.

Effective use of energy diplomacy can also help diversify both fuels and their sources of supply under long-term arrangements, thereby raising energy security. Such initiatives include transnational pipelines, equity oil and gas, equity coal, and energy plantations in Latin America and Africa wherein some countries have a marginal advantage in the production of biofuels.

Implement an integrated renewable energy policy

The role of traditional fuels in India's energy mix cannot be overlooked when considering India's energy supply options. Traditional biomass will remain the third most important energy resource of India after coal and oil, even in 2031–32. It is projected to meet 10–12 per cent of India's total primary energy consumption in 2031–32. This could prove to be an underestimate if agricultural growth of 4 per cent remains essential to India's ability to deliver an eight per cent + growth over the next 25 years. This level of agricultural growth would certainly raise the share of biomass, a carbon-neutral energy resource, in India's energy mix to levels higher than the foregoing estimates.

Using India's land resource offers many energy alternatives and choices. The theoretical potential of bioenergy is bigger by far than hydro, wind or nuclear. In fact, under certain assumptions bio energy in India could potentially deliver some 2.5 times the combined potential of all three together by 2031–32. Assuming India could actually find 60 million hectares of land for energy plantations, commercial wood plantations delivering the best-known yields of 20 tons of wood per annum per hectare could meet 29–35 per cent of India's primary energy requirement in 2031–32 (see Planning Commission 2006 for more). If currently used agricultural, forest and animal waste, wood chips and so on are included, proponents of bioenergy have argued that, theoretically bioenergy could meet a significant amount of India's projected primary energy needs in 2031–32. However, such bioenergy potential cannot be realized by using only fallow land or wasteland or non-arable land. Getting the best-demonstrated yields will require intensive cultivation, with water, fertilizer and pesticides.

To put this in perspective, just 2.25 million hectares of land under solar cells with 15 per cent conversion efficiency could yield the same energy as 60 million hectares of wood plantations. And seven to eight million hectares of land under solar cells could give India total energy independence in 2031–32. Solar cells can be used in arid land, deserts, mountaintops, and on home or vehicle roofs. They do not need fertilizer or pesticides, and use water only for cleaning collector surfaces. Furthermore, there is a growing body of evidence that 'green' fuels currently in use are not truly green and that the potential impact of large-scale energy plantations or energy crops on food security is critical.

Production and consumption of bioenergy in a localized and decentralized manner – consistent with age-old patterns – is indeed sustainable. However, it is far from clear what adaptive measures are needed for large-scale commercialization of these fuels so as to make commercial energy plantations and energy crops sustainable. Data on overall energy balances and the potential impact on global and local ecosystems, socioeconomic settings, local lifestyles, livelihoods of indigenous people, migration, land holdings, food security, and water security need to be established and authenticated.

Research is also required to establish viable germplasms and genotypes for bioenergy. The current research gaps are indeed immense, and socioeconomic viabilities are far from certain, and, typically, area- or region-specific. In any event, a combination of energy plantations, energy crops, wind, small and micro hydel, biomass-based co-generation, and solar could deliver at least five per cent of India's energy needs by 2031–32.

As far as large scale hydroelectric power is concerned, exploiting the full hydro potential of 150,000 MW, even though its contribution to the energy mix remains small at two per cent, remains a high priority because of its ability to provide peak energy as well as water storage capacity, which remains extremely low in India by global standards. However, hydropower has a number of environmental, governance and political hurdles to overcome. Apart from the issues of submergence, loss of forests and loss of biodiversity, there are many issues related to rehabilitation and the sharing of waters that dog hydro projects and delay their realization.

Pursue an informed nuclear policy

Expanding the use of nuclear energy for civilian use through international cooperation, and speeding up the development of domestic fast-breeder technology to ultimately enable the use of domestic thorium, remain national priorities that could potentially deliver up to five per cent of the energy mix by 2031–32 under the most assumptions for this sector.

Among the various options available to address India's energy needs, the nuclear option raises the most passionate debate; encompassing the social, economic and political dimensions of sustaining energy and climate security. Irrespective of which side of the political spectrum one is on, one cannot deny that the nuclear debate could definitely benefit from greater transparency, disclosure and sharing of available knowledge in respect of uranium reserves and their likely civil nuclear power potential; the risk of weapon proliferation and the risk of fissile material reaching wrong hands as a result of an expanded civil nuclear regime; the likely contribution of nuclear power in GHG abatement; public perceptions of safety, environmental and health concerns, as relevant to an expanded civil nuclear program; and the economics of nuclear power, including long-term management of nuclear waste. While these five concerns are as real as the nuclear option itself, there is no justification for excluding nuclear, or, for that matter, any other currently available option that offers energy and climate security. The simple truth is that there is no single silver bullet that ensures the energy and climate security of our planet and one must pursue and research all available options unless there is overwhelming and conclusive evidence to the contrary. Sustainability is all about creating options, not ruling them out.

Market reforms and risk management

A critical ingredient in all energy supply options is the existence of efficient and competitive upstream and downstream energy markets with minimal price and tax distortions and low entry barriers for both domestic and international players. Thus creating political consensus around the fact that India's energy security would indeed be boosted by roping a couple of global oil majors into the Indian market, even through privatization of some State oil companies, is an essential element of a strategy aimed at securing India's energy supply options. Such a move would strengthen domestic energy markets and integrate them with global energy markets.

Another element of a policy for a sound energy supply chain is to prepare for disruptions, which are likely, through the strategic storage of oil and gas. Such strategic oil and gas reserves can address the risks of technical and supply disruption in addition to providing a tool for managing global price volatility. Capturing part of existing commercial storage as a strategic reserve; buying options on existing storage in countries such as Singapore; and pooling India's strategic reserves with, say, European reserves, are all elements of far-sighted management of the energy supply chain.

Technology acquisition

Acquisition of efficient commercial and near-commercial technologies that yield higher efficiencies in energy extraction, conversion, transportation, distribution, storage and end use can significantly increase India's ability to meet its future energy needs. National technology missions that seek to expand India's domestic energy resource base through in-situ coal gasification, clean coal technologies, bioenergy, solar energy, thorium, and gas hydrates are all necessary for ensuring India's ability to meet its energy needs. And finally, participation in global efforts such as those in the field of fusion, zero emissions and carbon sequestration should also be pursued as an effective response to meeting India's modest, yet significant energy needs.

Conclusion

India's ability to deliver an annualized growth of eight per cent + over the next 25 years or so is critically dependent on its ability to reliably secure and supply the modest levels of energy essential to firing this growth. Making this growth inclusive, ensuring all-round improvement in human development parameters and meeting the millennium development goals require that India delivers lifeline levels of modern, commercial energy to all her citizens. This is a gigantic challenge in itself and meeting this challenge is fraught with threats, such as availability at affordable prices, environmental constraints and the unavailability of necessary energy infrastructure. Efficient and sustainable use of energy in the broadest sense will make the challenge more manageable and increase India's ability to deal with the potential threats. Finally, integrating India's response to the management of energy supply with global technological, institutional and market developments in the field of energy would go a long way in improving India's energy security.

Notes

1 Updated version of the paper delivered at the TERI-KAF Conference on Energy Security September 29, 2006.
2 Unless stated otherwise, forecasts in this chapter are based upon the 2006 Integrated Energy Policy report (Planning Commission 2006).

References

Planning Commission, 2006. *Integrated energy policy*: Report of the Expert Committee, Government of India.

UNDP, 2000, *World Energy Assessment: Energy and the Challenge of Sustainability*, New York: United Nations Development Programme, pp. 112

3 Energy and poverty in India[1]

Eshita Gupta and Anant Sudarshan

The notion of poverty is multidimensional. It is associated with numerous inter-related deprivations – lack of access to adequate levels of income, food, water, clothing, shelter, sanitation, health care and education. Although a definition of poverty normally does not directly include energy as a component, access to good quality energy sources is a critical determinant of the various dimensions of poverty. In accordance with Sen's capability framework, energy carriers can be perceived as commodities that outline an individual's capability set and thus facilitate his functioning in society (Andreas 2006). At the most fundamental level, energy delivers cooked food, clean water, health, space conditioning, and education services.

A number of studies have defined 'energy poverty' as the absence of sufficient means of accessing adequate, affordable, reliable, high quality, safe and environmentally benign energy services to support economic and human development (Reddy 2000, Barnett 2000, World Bank and UNDP 2005 etc). People who are considered 'energy poor' often rely heavily on traditional biomass fuels for cooking and simultaneously lack access to even minimal electrical lighting. As per IEA (2007) and UNDP (2007–08) statistics there are still about 2.5 billion people who continue to rely heavily on traditional cooking fuels and about 1.6 billion who have no access to electricity in developing countries.

India, home to more than a quarter of the world's poor, alone accounts for approximately 50 per cent of energy-impoverished people who have a high dependence on traditional cooking fuels, and for 31 per cent of people without access to electricity (IEA 2007, UNDP 2007–08). The dependence on biomass for cooking and heating causes more than 400,000 premature deaths (mostly women and children) in India annually (IEA 2007).

Apart from high dependence on traditional fuels, the vast majority of poor people in India have very low levels of energy consumption. In 2000–01, two-thirds of the population used less than 20 GJ of primary energy per capita. This is a third of the global average of 2 kilowatts per capita and less than the minimal amount of 1 kilowatt per capita estimated by Goldenberg (1990) as necessary to meet basic needs (Pachauri 2007).

The purpose of this chapter is to provide a brief overview of the extent and the nature of energy poverty (as measured by poor access to clean energy) in India. We analyze the present situation with respect to energy access based on the National Sample Survey (NSS) 2004–05 datasets, which provide an excellent means of understanding the distribution of households by the primary source of energy used for cooking and lighting. The data highlight the fact that access to modern cooking fuels is severely inadequate in rural areas. Firewood is still the dominant fuel in the rural sector even in the top expenditure groups. Many rural poor also now depend on purchased firewood (as a result of firewood's decreasing availability) although this trend is more common among rural rich and urban poor. It is evident that the use of electricity is still quite strongly associated with the relatively well off in India, especially in rural areas. While in urban areas the majority of households across income deciles use electricity for lighting, in rural areas this is the case only among the top six income deciles.

We observe that disparities and inequalities in access to modern energy are growing between urban and rural households, between different geographical regions, and between the top and bottom expenditure deciles of the population. While access to electricity and LPG is increasing in both the rural and urban sectors, the rate of penetration is very slow in the case of the rural sector. The major constraints faced by rural people are their low purchasing power and the poor availability of these modern fuels in remote rural areas. Universal LPG subsidies have been regressive, benefiting the rich far more than the poor. Large amounts of kerosene are diverted to the automobile sector. Even in villages connected to the grid, electricity supplies are highly irregular. There has been a signal failure to ensure lifeline supplies to really poor people in rural India. In that context, the success of any energy programme and government policy needs to be gauged by the extent to which it actually benefits the poor.

The structure of this discussion is as follows. We begin by reviewing the literature on the energy-poverty problem. We then discuss patterns of household energy use in India over the period 1993–94 to 2004–05 using NSS household-level energy data. Thereafter, we highlight major constraints faced by households in switching from traditional to modern fuels. Finally we draw major policy conclusions and look at a few lessons learned.

Energy-poverty problem: literature review

Poverty and 'energy poverty' have numerous interactions. Recognizing this fact, we begin by highlighting the characteristics of the so called 'energy poor' or 'energy vulnerable'. Most of the poor have limited livelihood opportunities, poor land and lack a regular cash flow. Apart from income poverty, they often lack basic education, health and other social services. Their lesser ability to pay for modern fuels and energy-using equipment results in a natural preference for freely available, but inefficient, traditional fuels. Even when improved energy is affordable to the energy poor, it is often not easily accessible in remote areas. It is generally financially unviable for state utilities to construct capital-intensive distribution networks to meet the energy demands of widely scattered low energy-consuming communities. Thus, poor people tend to have a much smaller range of options than the rich when it comes to determining which fuels to use and what equipment to buy (Ramani and Heijndermans 2003).

Consequently, the poor find themselves using traditional dirty fuels (such as firewood, dung and crop residues) in inefficient devices (with as low as 8–10 per cent efficiency), that provide little useful energy. In addition they expose themselves to serious health risks. This is particularly true for women and children, who are disproportionately exposed to indoor pollution resulting from the combustion of solid fuels. A number of studies have highlighted, besides the health risks, the significant time costs associated with the collection of these fuels (Smith and Mehta 2000; UNDP/ESMAP 2003; Parikh *et al.* (2001, 2003, 2005); ESMAP 2002; Dutta 2005). Parikh *et al.* (2005) found through a household survey of four Indian states that 85 million households spent 30 billion hours annually just in gathering fuel wood; and that 24 million adults had respiratory symptoms. The total economic burden of using dirty biomass fuels was estimated at about 300 billion rupees, using a wage rate of Rs 60 a day and accounting for opportunity costs, time off work due to health reasons, and direct medical costs of respiratory and eye diseases. A close link was found between the negative impacts of biomass and gender, since it is women who spend time collecting the fuel and who are exposed the most to combustion byproducts while cooking.

In addition to severe costs of health and time, being unable to use modern energy and equipment results in lower productivity and poor quality output, and thus, consequently, lower incomes from economic activities such as agricultural farming. This leads to lower returns for labour and invested capital, and consequently a smaller surplus for reinvestment, resulting in the so called 'energy poverty trap'.

There are numerous studies that describe the impact of improved access to modern fuels on the living conditions, lifestyles and livelihood enhancements of the poor – all important for enabling people to escape the 'poverty trap' (Smith and Mehta 2000; UNDP/ESMAP 2003; Parikh *et al.* (2001, 2003, 2005); ESMAP 2002, Andreas 2006). Kemmler and Spreng (2007), using data from the National Sample Survey Organization (NSSO) found very high correlations between low levels of energy consumption (defined in terms of access-adjusted useful energy) and various poverty measures such as illiteracy, lower levels of consumption expenditures, small size of landholdings, poor housing conditions and so on.

Further, as a result of increasing deforestation, there is a growing commercialization of firewood even in rural India, increasing the vulnerability of the 'energy poor' to energy price shocks. Consequently, many poor people pay cash for commercialized firewood (due to the scarcity of firewood) and to some extent for highly priced electricity (required to meet basic needs). They usually end up paying more than those who are well off, both in absolute terms (price per unit of useful energy) and relatively speaking, in that they spend a much larger portion of their low income on fuel and light (ESMAP 2002; UNDP/ESMAP 2003; Ramani and Heijndermans 2003).

Another branch of the literature relates to the numerous determinants of fuel switching (from traditional to modern fuels). A number of studies in India and many other developing countries have shown that fuel choices depend on a complex set of factors, where monetary costs, gender issues, ease of availability, capital costs, and cultural preferences may all combine to encourage the persistent use of traditional biomass fuels over more energetically efficient modern alternatives (Table 3.1).

In a number of field surveys it has been found that households, particularly in rural areas, even as they climb the income ladder, do not leave out traditional fuels (Heltberg 2004, Masera *et al.* 2000, World Bank 2003). They adapt to modern fuels like LPG but continue using fuel wood. Once modern equipment is acquired, devices using fuels down the ladder such as wood are used as insurance against supply failure. Irregularities in the supply of modern fuels like LPG, high initial fixed costs, and maintenance costs, are prime factors that hinder a complete and smooth shift towards the use of modern fuels from traditional ones.

A household energy survey of Delhi conducted by TERI shows that most poor households in Delhi use a mix of LPG, kerosene and biomass for cooking. Most slum dwellers relied on kerosene and biomass as their primary fuels for cooking due to an inability to get legal connections as well as the high upfront costs of LPG compared to other fuels. Relatively affluent households in the slums used LPG as their primary fuel for cooking, but resorted to the use of biomass and kerosene as a backup fuel during periods of delay in refilling the LPG cylinders or during a shortage of the fuel (TERI 2007). Vishwanathan and Kumar (2005) found, in a study of the use of cooking fuel in Indian households, that rural households across income groups used fuel wood as their primary energy source when it was easily available. In a few states there was some evidence of shifts down the energy ladder – from fuel wood to crop residues – due to easier access.

Foley (1995) has argued that, as incomes rise, additional energy services are required and these services often demand modern sources of energy. Traditional fuels may continue to be used for the most basic energy services (cooking and heating) while modern forms such as electricity may provide other services (such as lighting, radios, televisions, refrigerators). Thus it is not necessarily the case that *fuels* form a hierarchy (as in the energy ladder model), but rather that demand grows hierarchically for different *services* as incomes grow.

The characteristics and vulnerabilities of the energy poor place them at a distinct disadvantage compared to the rest of the population. There exists a growing literature that highlights growing energy inequities between the poor and non-poor in India. Unsurprisingly, the poor consume significantly less energy in absolute terms and use very different technologies compared to the better off. For instance, the poor with limited access and affordability use minimum electricity for lighting and are generally unable to take advantage of productivity enhancements (such as electricity-based agricultural production and micro-enterprises) that are necessary to break the vicious circle of poverty (ESMAP 2002; UNDP/ESMAP 2003; Ramani and Heijndermans 2003).

Pachauri *et al.* (2004) constructed a two-dimensional energy poverty measure for India that combined access to different energy carriers with the quantity of energy consumed per capita. They found that the number of people living in extremely energy-poor households in India, having access to just biomass and kerosene (and often only just enough to cook a full meal a day), decreased from 38 per cent in

Table 3.1 Factors influencing energy choice

Micro-level characteristics influencing household energy choices	
Household income	Dominant variable determining fuel choices (e.g. ESMAP 2003, Leach 1992, Jiang 2004, Pachauri 2004).
Availability	Availability is important because studies suggest that energy security, as well as cost, determines household fuel choices (e.g. Soussan 1989). Fuel availability in turn depends on distance to markets, distribution systems and local resource endowments.
Distribution of costs/capital costs	Low-income households are often unable to purchase the equipment necessary to use energy in the form of electricity/LPG (appropriate stoves, LPG cylinders, expense of wiring house etc.). Thus high initial capital costs are a significant barrier, even if life-cycle costs and marginal costs of modern energy forms are lower than for traditional energy.
Energy technology and support systems	Technology that is affordable, does not require large lifestyle changes, is adapted to local conditions, is easily available, and has the requisite maintenance, training and financing systems to be usable, is a prerequisite to successful fuel transitions. Often these aspects are not fully taken care of, leading to the failure of initiatives such as biogas plants or improved cook stoves.
Climate	Energy use in colder areas tends to outstrip that in warmer areas. Energy for heating (often through wood combustion) is a basic energy service, while air-conditioning and even fans tend to come into use only at higher income levels.
Fuel prices	High commercial, modern fuel prices reinforce the perception of traditional fuels being much cheaper (essentially 'free'). At the same time, even in urban areas where traditional fuels are also bought and sold, the price of modern alternatives acts as a ceiling for the price of options such as fuel wood.
Household size	Larger households need more energy but also achieve economies of scale and may have higher incomes. The playoff between these factors makes household size important in determining the extent of energy use and the nature of the fuel mix.
Gender roles	In rural areas in particular, women bear most of the health and time costs of using traditional fuels. However, decision-making powers are primarily vested in men. This asymmetry between who determines whether to purchase commercial fuels, and who is most affected by them, can greatly slow down a transition. This is particularly the case when the opportunity cost of the woman's time is low.
Wage rates	Wage rates and employment opportunities, particularly for women, affect both overall household income and the opportunity costs of time spent cooking food or collecting traditional fuels. If wages for work outside the house are high enough, it makes economic sense for a household to switch to more efficient fuels and appliances.
Cultural preferences	Traditional fuels may persist due to cultural preferences of different kinds. For example, some foods, such as traditional breads, taste different when cooked on a modern cooking appliance and households may continue to use a traditional stove to prepare these food items.

1983–04 to 14 per cent in 1999–2000, but the inequities in distribution of energy consumption and access increased significantly over the same period. For instance, in rural India just one third of the extremely energy poor were able to improve their situations as against the corresponding fraction in urban areas. Further, while there was a decline in the number of very energy poor in the bottom energy consumption segment, there was only five per cent reduction in the total number of people in the bottom and lower consumption segments taken together. At the same time the number of people in three relatively better off segments (high consumption consumers having access to either electricity, or LPG, or both, along with other fuels) increased from three million in 1983 to 160 million in 1999.

Similarly, Vishwanathan and Kumar (2005) and ESMAP (2002), undertook an analysis of cooking fuel use in India for different states between 1983 and 2000, based on NSS household-level data. They found marked differences in the extent of transition to modern fuels in different states, as well as differences in the progress of the energy transition across income groups and urban and rural areas.

Patterns of household energy use in India

We discuss here the NSS sixty-first round (July 2004–June 2005) survey along with a comparative picture of earlier periods i.e. the last two quinquennial rounds (fiftieth (1993–94), fifty-fifth (1999–2000)) of household consumption expenditure.

Patterns of energy use for cooking

The major cooking fuels in India are firewood and chips, dung, kerosene and LPG. According to the NSSO 2004–05 survey 124.98 million households (about 60.57 per cent) used firewood and chips as a primary source of energy for cooking. About 7 per cent of households used dung, 3.7 per cent used kerosene and 22 per cent used LPG.

In rural India, households primarily use three sources of energy for cooking –firewood and chips, dung cakes, and LPG. It is observed that a more or less constant percentage (75 to 78 per cent over the period 1993–2004) of rural households are dependent on firewood and chips, due to their cheapness and easy availability in the rural sector. It was observed that on average, 46 per cent of households using firewood and chips in rural India obtain these fuels through 'free' collection; about 21.14 per cent of households depend on home-grown stock; and 23.7 per cent make cash purchases. Whereas the use of LPG has gradually increased in rural India from 1.9 per cent in 1993–94 to 8.6 per cent in 2004–05 owing to its improved availability and convenience, the use of dung cakes has declined only marginally from 11.5 per cent in 1993–94 to 9.1 per cent in 2004–05. This slow drop provides some cause for concern.

In urban India, LPG is the most commonly used primary cooking fuel, followed by firewood and chips, and kerosene. More and more households have switched from firewood and chips and kerosene to LPG. In 2004–05, 57.1 per cent of urban households met their primary cooking needs by using LPG as compared to just 29.5 per cent in 1993–94. Interestingly, LPG may have displaced kerosene more than it has displaced firewood. The decline in the use of kerosene, from 23.2 per cent to 10.2 per cent is greater than the decline in the use of firewood and chips, from 30 per cent to 21.7 per cent. Firewood is the second most commonly used primary cooking fuel, with more than one-fifth of the population dependent on it as a primary source of cooking energy. Unlike in rural areas however, it is a commercialized fuel in urban India, with about two-thirds of urban firewood-using households having purchased the fuel.

Further, one observes that many households use multiple fuels for cooking. This means it is necessary to be very careful when interpreting data on the spread of any single fuel in isolation. For instance, in 2004–05, about 86 per cent of rural households used firewood and chips, as against 75 per cent using it as the primary source discussed previously. Thus the adoption of a modern fuel has not meant that the traditional fuel disappears entirely. Similarly, LPG was used by about 11.7 per cent of rural households, as against 8.6 per cent primary users. Only 2.7 percent of the LPG users were using it as the only cooking fuel. All others were multiple fuel users.

Access differences based on income

In order to study income-related differences in the use of fuels, we further disaggregate data by income deciles. Households in both the rural and urban sectors are arrayed in order of increasing per capita expenditure and categorized into 10 groups – by monthly per capita expenditure deciles – each having the same number of households. The first decile (D1) refers to the poorest 10 per cent of the population while the 10th decile (D10) corresponds to the richest 10 per cent of the population. Figure 3.1 gives the proportion of households using firewood and chips, dung, kerosene and LPG as the primary source of energy for cooking as a function of per capita expenditure decile, both in rural and urban India.

The graphs in Figure 3.1 indicate that while LPG tends to be a normal good with higher income resulting in higher consumption, solid fuels (firewood and chips, dung) behave largely as an inferior good with consumption declining with income. This is in accordance with the energy ladder of fuels, with people climbing the ladder from traditional to modern fuels as their purchasing power increases (see Leach 1992; Barnes and Floor 1996).

In both the rural and the urban sectors, a larger number of households belonging to the lower deciles used more firewood and dung and less kerosene and LPG compared to households in the higher MPCE deciles. Figure 3.1 clearly shows that the use of the two traditional fuels (firewood and chips, dung) declines very sharply in the urban sector, whereas in the rural sector the use declines relatively less steeply. In fact, monthly per capita consumption of firewood and chips increased in absolute terms from 17.47 kg in the lowest income decile, to 22.7 kg in the highest income decile, in the rural sector. In the urban sector, it is much harder to access traditional fuels and even fuel wood needs to be purchased, so, with higher per capita incomes, cleaner alternatives such as kerosene or LPG become attractive

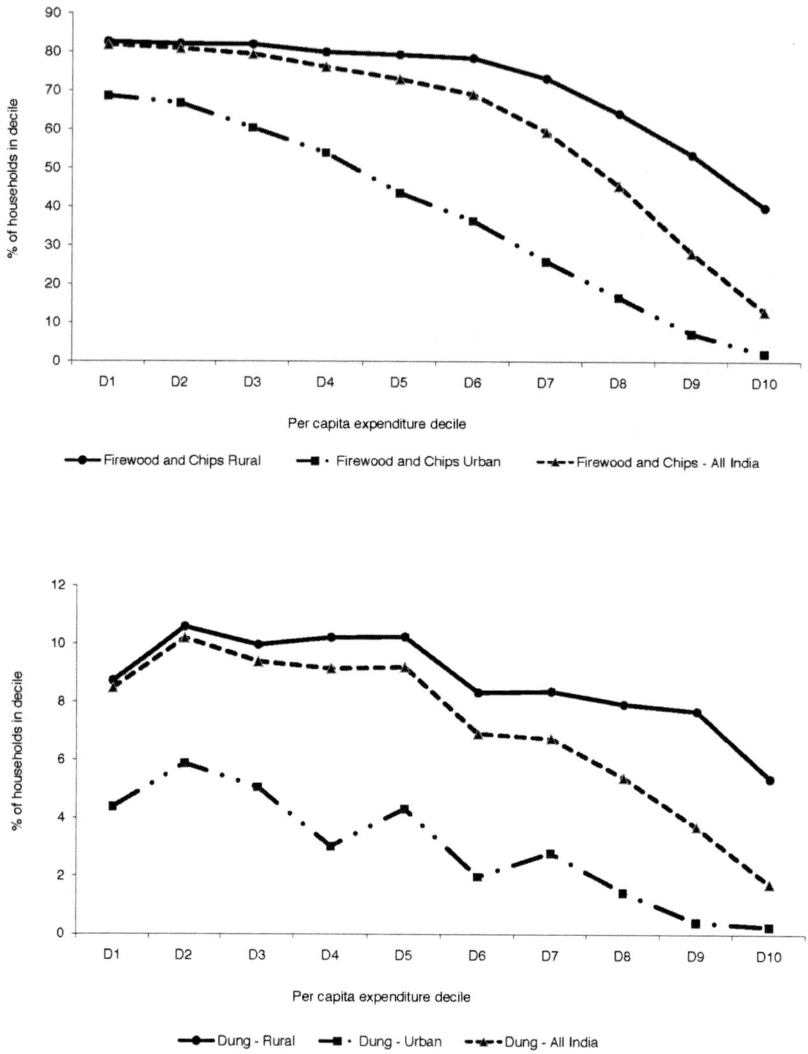

Figure 3.1 Primary cooking fuels by MPCE deciles in rural and urban India (2004–05).

options. In contrast, in the rural sector, high fuel prices reinforce the perception of traditional fuels being much cheaper (essentially 'free'). The high dependence on traditional fuels even among higher income groups in rural India can partially be explained by the fact that they generally possess more land and livestock, and thus tend to have easier access to freely available home-grown stock. However, when richer people depend on biomass they usually employ hired help to collect it (Leach 1992). Also, they are able to purchase more efficient biomass stoves (even available with chimneys) reducing their exposure to indoor pollution.

The use of kerosene is miniscule in the rural sector and rises consistently with income deciles, yet even in the top two deciles, just 3 per cent of all households used kerosene as a primary source of energy for cooking. The fact that kerosene is a rationed fuel may have discouraged households from adopting it for cooking so that its limited supply could be used for other functions such as lighting (perhaps as a backup source). In the urban sector, on the other hand, with its easy availability through public distribution systems, kerosene has emerged as a first transition fuel for the urban poor, with its use first increasing to exceed 15.12 per cent by the 6th decile and then declining to reach 6.14 per cent in D10 (ESMAP 2002; UNDP/ESMAP 2003).

The use of LPG in the urban sector increases substantially with rising per capita expenditure, being used in more than 50 per cent of households by decile 7 and 80 per cent of households by decile 10. In rural India, a very small proportion of households belonging to the bottom segments used LPG as a primary source of energy. It is observed that the use of LPG increased sharply in the top three deciles, to reach 42.81 per cent in D10 (still just half of the consumption in D10 of the urban sector) from a small 0.19 per cent in D1. Clearly, LPG penetration is still very low in rural regions and almost nonexistent amongst the poorest households. While the energy poor in the urban sector have the choice of using LPG (with readily available supply), their inability to get legal connections, as well as the high upfront costs of LPG (compared to other fuels) are the major factors for their not switching to modern fuels. In the rural sector, on the other hand, apart from lower affordability poor households also suffer from inadequate supply. This is reflected by much lower levels of adoption, even by the richest 10 per cent of the rural population.

Access based on occupation

NSS data reveals that household energy use is also influenced by the occupation groups of the householders – defined as agricultural labour, non-agricultural labour, self employed in agriculture or non-agriculture, other labour, other households (having very diversified income sources), regular wage earners or casual workers, and so on. In the rural sector, the use of solid fuels (wood and dung) was highest among agricultural labourers (85 per cent). Agricultural labourers usually live in very poor conditions, have lower opportunity costs of time, and thus depend largely on freely available fuels. In the urban sector, the use of solid fuels was highest among casual labourers (58 per cent) and lowest among regular wage earners (12 per cent). Higher wages in the urban sector make it economically sensible for households to switch to more efficient fuels and appliances.

Access based on social classification

In rural areas, the schedule tribes (90 per cent) and schedule castes (77.2 per cent) use significantly larger amounts of traditional fuels as compared to other classes. Although these social groups usually live in poor and backward areas, this may also suggest the existence of access discrimination (like rationed kerosene). In urban sectors, LPG is the most commonly used fuel across all social groups.

Access based on geographical location

Statistics also show significant geographical variations in patterns of use. In rural India, LPG has the highest penetration in the three high income states of Punjab (24 per cent), Haryana (19 per cent), and Kerala (18 per cent). Similarly, in the urban sector, Punjab (70.3 per cent), Haryana (72.9 per cent) and Gujarat (62.3 per cent) are amongst the highest users of LPG. Although relatively poorer states largely have a lower penetration of modern fuels (such as Bihar, Orissa, Chhattisgarh, Assam and Jharkhand) one observes an absence of absolute linear relationships between a state's income and the fuel mix. For instance, Kerala (urban), one of the most developed states, is observed to have a much higher dependence on firewood and chips (48.4 per cent) and a corresponding lower LPG penetration (43.7 per cent) as compared to Assam (urban: 27.2 per cent firewood and chips; 60.6 per cent LPG) and Rajasthan (urban: 38.6 per cent firewood and chips; 51.3 LPG). This is reflective of the fact that Kerala is a heavily forested state where wood is readily available. Similarly, in high-income rural Punjab, where dairy farming is widespread, a large number of households (33.3 per cent) use dung as a primary cooking fuel. Thus, easy availability is an important determinant of traditional fuel use.

Patterns of energy use for lighting

In India, kerosene and electricity are the two major sources of energy for lighting, accounting for about 99 per cent of households in both rural and urban areas[2]. Over time, electricity is increasingly displacing kerosene as a source of lighting both in rural and urban India. Between 1993–94 and 2004–05 about 17 per cent of households in rural India shifted from kerosene to electricity, as against 9 per cent in the case of urban India.

In the rural sector, the use of kerosene as a primary source of lighting is very high; 44 per cent of rural households were dependent on kerosene in the year 2004–05 as compared to just 7 per cent in urban India. In the same year, 54.9 per cent of rural households derived their lighting energy requirements from electricity as against 92.3 per cent in urban India.

Many households with access to electricity exhibit significant consumption of kerosene for lighting, particularly in the rural sector. This shows that such households may be using kerosene-based lamps in case of electricity supply failures; and that even households connected to electricity have not necessarily made the transition away from kerosene completely – whether it is used as a lighting backup fuel or as a supply source to run a small generator for times when the mains supply is unavailable.

Access based on income

Figure 3.2 shows the proportion of households using kerosene and electricity as the primary source of energy for lighting as a function of per capita expenditure deciles both in rural and urban India.

In rural India we observe that the majority of the households between decile 4 and decile 10 used electricity as the major source of energy for lighting. However, among lower MPCE groups the use of kerosene is still very high, with almost

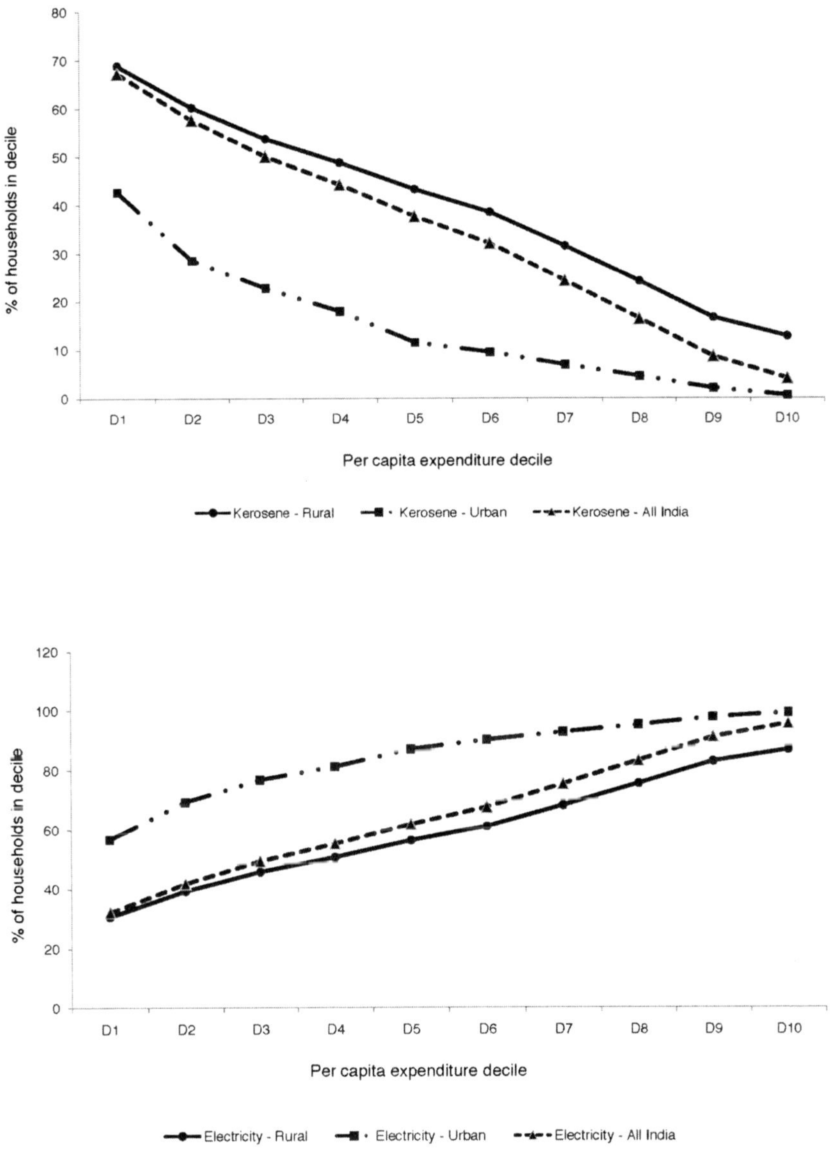

Figure 3.2 Households using kerosene and electricity as the primary source of energy for lighting.

70 per cent of households in decile 1 depending on kerosene in to meet their primary lighting requirements. The high dependence on kerosene among low-income households is explained by the high installation cost of electricity services and equipment; poor-quality housing; and, at the same time, a much greater preference for mobile lighting, which is better achieved with kerosene lamps.

In urban India on the other hand, access to electricity depends less on income as the majority of households use electricity in all income deciles with the penetration reaching over 90 per cent by decile 6 from 56.6 per cent in decile 1. In urban India the difference in electricity access between richest and poorest appears to be much less as compared to the rural sector.

Access based on occupation

Again, in rural India, access to electricity is found to be least among the agricultural labourers (47.6 per cent). In urban India, the use of electricity is highest among regular salary earners (97 per cent) and lowest among casual labourers (76 per cent).

Access based on social groups

In both rural and urban India, the use of electricity was less frequently observed for scheduled tribes (42.7 per cent in rural; 83.8 per cent in urban), followed by scheduled castes, and then by other backward classes and 'others' (64.3 per cent in rural; 96.1 per cent in urban).

Access based on geographical location

There are significant geographical differences in electrification. The use of electricity is found to be positively correlated with the level of income in different states. In rural India, the use of electricity was highest in Punjab (96 per cent of the households) followed by Haryana (90 per cent), Karnataka (86 per cent), Tamil Nadu (84.6 per cent) and Andhra Pradesh (84 per cent). The proportion of households using electricity is lowest in Bihar (10 per cent), Uttar Pradesh (24 per cent), Jharkhand (26 per cent), Assam (30 per cent) and Orissa (31.5 per cent). These are also the states with the greatest number of villages not connected to electricity. Kerosene is still used predominantly in these states. One also observes some correlation between the state economy's dependence on agricultural land and irrigated areas, and rural electrification. Apart from Punjab, Haryana and two middle-income states, namely, Andhra Pradesh, and Maharashtra (with a large net sown area, and large areas of irrigated land) have high rural electrification, because electricity is extensively used in agricultural pumping and cropping.

In urban India, in 10 out of 17 major states, over 90 per cent of the households obtain their lighting energy requirements from electricity. In the urban areas, the percentage of households using electricity is again found to be lowest among the two poorest states –Bihar (68 per cent) and Orissa (70 per cent).

Major constraints on the transition to modern fuels

Trends in energy use show that neither growth in income growth nor the affordability of alternatives can alone bring about the change to using modern fuels. There are numerous other factors, such as fuel availability, culture, geography, the extent of urbanization, fuel prices and the design of state policies that influence fuel choice. This section seeks to describe some of the major constraints faced by households in the process of undertaking the energy transitions just discussed.

Part of the reason the use of traditional fuels has persisted is that the Indian Government has been unable to provide affordable, abundant and highly reliable modern alternatives. Present energy policies tend to have an urban bias, with much focus on commercial fuels – coal, oil and gas – fuelling urban energy consumption and growth. Further, there has been very little integrated planning by the state. Both market-driven and subsidy programmes have run side by side, which is counterproductive. Typical of many government efforts, targets and incentives in different schemes are set independently of each other. Individual targets for the improved chulha programme and biogas plant projects have sometimes been known to cumulatively exceed estimated demand (TERI 2007, ESMAP 2003). Some of the major problems in this context are discussed below.

Poor targeting of subsidies and programmes

In the past, subsidies have been largely mis-targeted. Broad-based LPG subsidies, perversely, tend to benefit the better off much more than the truly poor, because of the greater consumption and greater ease of access of those who are wealthier (Saghir 2004, World Bank 1996, Barnes and Floor 1996, Pitt 1985). A World Bank study highlighted that in 1999–2000 three-quarters of the LPG subsidy went to urban households, four-fifths of which was given to the top 50 per cent of the population (UNDP/ESMAP 2003). High subsidies also make it hard for utilities to stay in business, upgrade equipment and expand services. In the long run, they are often financially unsustainable. Subsidies with quantity rationing often lead to corruption and problems of access.

In the case of kerosene, a significant fraction of the subsidised fuel is illegally diverted and used to adulterate diesel (a practice encouraged by the price differential between diesel and kerosene). Additionally, the remaining kerosene that does reach consumers suffers from the same problems of poor targeting as plagues LPG subsidies (UNDP/ESMAP 2003).Most electrification programmes focussed more on the number of villages connected to electricity than households. Many of the specifically targeted schemes (such as *kutir jyoti*) for people below the poverty line have not been successful due to large scale leakages and uniform flat rates regardless of the level of consumption. Excessive subsidization has made the utilities financially unviable. Due to increasing financial constraints there have been providers of utilities reluctant to promote such schemes as would benefit the poor (ESMAP 2002, Ailawadi and Bhattacharyya 2006; Modi 2005).

Subsidies have meant that LPG supplies are limited and the electricity supply, even where grid connections are available, is highly unreliable. Similarly, while kerosene has many advantages over fuels such as firewood, it too needs to be subsidized. Extending energy sources such as LPG or electricity to India's vast rural population will necessitate large increases in the imports of fossil fuels. With the need to find long-term, credible solutions to concerns about the country's energy security, chances of complete elimination of traditional fuels in the foreseeable future are very small. The use of traditional sources is not undesirable in itself–it is the manner in which they are used that is the problem. As fuel prices rise and global energy supplies become scarcer and more threatened by geopolitical tensions, this is a huge problem for India to deal with. It is also a problem that the country cannot afford to ignore – as we have pointed out, the costs of energy poverty are extremely high. To tackle these twin problems, an approach that has been tried in India has been to encourage the use of clean decentralized renewables in rural India, as well as cleaner technologies that rely on traditional fuels. This approach has always had a great deal of potential – from the point of view of tackling both energy security and climate concerns. Unfortunately state efforts in this direction have had only limited success, especially in rural India. To understand why rural policies have not succeeded nearly as well as India would like, it is essential to appreciate the complexities of the energy poverty problem.

For example, an important factor holding back the spread of decentralized renewables and cleaner technologies, particularly in rural households, is that often the technology required is still relatively nascent and requiring further development. Unfortunately, options such as solar cookers, biogas plants and improved *chulhas* of various kinds, while promising technologies in the lab, have not turned out to be good substitutes for traditional methods in the field. Box solar cookers, for example, can be used to cook only food requiring certain cooking techniques (such as boiling). Therefore while it is possible to cook rice, pulses and boiled vegetables, it is impossible to fry food or make *rotis*. Newer parabolic cookers provide much greater heating but come with other problems. One drawback with these is that they are open cookers and consequently short spells of cloud and/or rain can completely ruin the meal. Biogas plants have tended to be a better-designed option, but require careful construction and regular maintenance. Problems such as construction quality, low gas output in colder climates and poor design for inputs other than dung are urgent issues that require research and development time (also TERI 2003).

Alternative technologies need to fit in to an existing 'energy ecosystem'. It is often forgotten that traditional energy sources or cooking tools, while perhaps 'inefficient', are part of a lifestyle that has adapted around them. Consequently alternatives need to be made close substitutes, or they must battle a resistance to changes in habits. One good example is the traditional wood-fired stove (*chulha*). This device is smoky and energy inefficient, but in practice has been found to have unexpected secondary uses. For instance, chillies or spices to be dried can be stored hanging over the *chulha*. Eliminating the smoke would mean creating the need to find a different way of smoking and drying some food items. In addition, in some areas, smoke from cooking serves the secondary purpose of keeping mosquitoes away. Similarly, it is common practice in some parts of India to cook either indoors or outdoors (depending on the season). Alternatives that are not portable will therefore need to battle ingrained habits to gain acceptance. Thus a new smokeless *chulha* with a chimney might end up being rejected by households, even though it might be more efficient and healthier.

Clean fuels tend to be perceived as losing out on the cost equation. A large number of rural households are characterized by low purchasing power, as well as access to fuels such as dung, crop residues or wood at no monetary cost. Gathering these fuels has significant time costs and using them to cook is a serious health hazard. However, the bulk of these costs are borne by women and children who possess relatively little negotiating power in the household. Decisions to spend money on biogas plants or new *chulhas* or LPG stoves are strongly influenced by men. Within the household, the value of a woman's time is regarded as extremely low and with the brunt of health costs not being borne by the person making purchasing decisions these are not factored in when making energy choices. Consequently, it becomes harder to convince many rural households of the value of paying for either clean technologies, or cleaner fuels such as LPG.

Institutional shortcomings and bad programme design and implementation have played no small part in the disappointing results from India's efforts to disseminate non-conventional energy sources. To take one example – the success or failure of these programmes has normally been judged by whether or not they meet prespecified targets. The targets have been defined in terms of units installed, with no reference to continued functionality after a period of time, backup services or actual use. Consequently, during implementation the thrust has been on pushing the programmes in as many areas as possible, backed by heavy subsidies, but without sufficient care being taken to ensure that there is a good match between the needs of users and the schemes being promoted.

Policy implications

The preceding sections have looked at the characteristics of energy poverty, complexity in the way households make energy choices, and major shortcomings in government programmes. An important question for policy makers is how to guide the fuel-switching process along a sustainable pathway, and how to ensure that it occurs as quickly and smoothly as possible. In this section we explore some of the issues that need to be kept in mind when framing policy with the objective of encouraging households to shift to newer and better forms of energy.

Addressing energy affordability

As poor households have low incomes and they spend a significant proportion of their income on fuels for cooking and lighting, the issue of energy affordability assumes significant importance. In principle smart LPG and kerosene subsidies specifically directed to the poor (perhaps implemented with energy coupons distributed to the poor) and financial support for the initial capital costs and non-recurrent costs can go a long way towards enabling poor households to make more efficient choices (Heltberg 2004, World Bank 2003).

Enabling widespread access

We have seen that affordability alone is not enough to motivate shifts in energy use. Enabling widespread access to modern alternatives is also crucial if transitions are to work. Unfortunately when fuel supply and distribution systems are controlled by the state (such as for kerosene or LPG in India), there tend to be inequities in distribution (with major urban areas being far better served than more remote widely scattered locations), black markets and corruption. Without making any comment on the advantages or disadvantages of state control over certain fuel markets, it is clearly the case that unless easy access can be ensured, equitable and widespread use is unlikely to be seen.

Another aspect of enabling easy access lies in making it possible for poor households to purchase small amounts of energy at a time. LPG for example, is sold in cylinders and therefore must be purchased in bulk. Given the reality of households with limited credit and limited willingness to pay large amounts at a time for energy, it is necessary to size the cylinder optimally for the consumer. However, the 'optimal' amount of fuel is clearly very different for an urban household with higher energy use and higher purchasing power, than for a poor, rural household. Thus even though buying large amounts of the fuel at a time reduces the average cost of energy, the initial capital costs of larger cylinders may well be beyond many households. As far as electricity is concerned, a similar problem occurs if low ampere connections are not available. In many cases rural consumers do not require connections rated for high loads, since they are not looking to run many appliances that might be used in a higher-income urban setting. Thus, ensuring the availability of electrical connections that take into account the needs of the consumer is important, and it is not necessarily the case that one size should fit all.

Appreciating the complexities governing energy choice

Large-scale programmes aimed at electrification or the provision of fuels such as LPG need to be framed after taking into account the fact that household energy choices are made on the basis of a number of factors, not just monetary costs. In addition, households often look to maximize their energy security and therefore prefer fuel mixtures. Thus merely setting up infrastructure does not guarantee a transition – actual use requires both consistent availability and the alignment of other factors such as capital costs, suitable technology, culture, and gender issues. All these issues need to be considered when determining what a realistic consumer response is likely to be, and what is the appropriate pricing and distribution mechanism that would allow the programme to be a success.

Taking cognizance of gender realities

As we have stressed earlier, energy poverty is linked quite closely with gender roles and the earning power and empowerment of women. Energy policy must therefore be framed with this fact in mind. It is crucial to involve women in the testing and design of new cooking technologies. For programmes concerning the use of improved cooking stoves or solar cookers, the end users are women, so it is important to convince them of the utility of the new technology. At the same time, men are often the decision makers when it comes to spending money on alternatives. Therefore it is important for both men and women to be a part of the technology dissemination process.

Changing societal awareness of the true costs of traditional fuels; increasing female participation in the decision making process; and getting the state to focus on this aspect of energy security are all prerequisites to substantive change.

Decentralization, modern renewables and sustainable biomass

Decentralized electricity generation, especially through the use of modern renewables such as solar photovoltaics and the sustainable use of biomass, provides options that should be at the forefront of energy policy in developing countries. While it is true that at this point in time the technology does not exist for these energy options to economically provide large-capacity additions, they still have an invaluable role to play in providing lifeline amounts of clean energy, especially in remote areas where grid electrification is more difficult. With the use of biofuels likely to continue for many years, even under optimistic projections, technologies such as wood gasifiers or biogas plants, and the spread of farm forestry and natural forest management, can make a huge difference towards reducing energy inequities and improving energy access.

Finally, we conclude by observing that energy developments cannot be made in isolation. Improved access to clean fuels will not result in human development by itself. The provision of adequate energy is a necessary but not a sufficient condition to shift the poor out of the vicious circle of 'poverty' or 'energy poverty'. This will occur most easily when other complementary inputs – employment, opportunities to use modern energy, education, connectivity and so on are all made available. It is easy to see why a remote household with poor employment options and little access to the fruits of development might consider the form of energy being used somewhat low on a wish-list of changes. Thus an energy transition is part of a package. While we may study it in isolation, it is far more likely that specific programmes will succeed if they come alongside other efforts or changes that help the process of development.

Notes

1 The authors gratefully acknowledge Mr Prabir Sengupta, Mr R K Batra and Dr Pradeep Dadich for their guidance and valuable comments on this paper. We are extremely thankful to M K Bineesan for his contribution for data and graphs.

2 In 2004–05, about 65.23 per cent of households (or 134.97 million households or 625.68 million individuals) met primary lighting requirements from electricity and 34.18 per cent of households (or 70 million households or 350.08 million individuals) met primary lighting requirements by kerosene.

References

Ailawadi, V. S., and Bhattacharyya, S. C., 2006. 'Access to energy services by the poor in India: current situation and need for alternative strategies.' *Natural Resource Forum*, (30), pp. 2–14.

Andreas, K., 2006. 'Regional disparities in electrification of India – do geographic factors matter?' CEPE Working Paper No. 51, Zurich: Centre for Energy Policy and Economics (CEPE).

Barnes, D. and Floor, W., 1996. 'Rural energy in developing countries: a challenge for economic development.' *Annual Review of Energy and the Environment* (21), pp. 497–530.

Barnett, A., 2000. 'Energy and the fight against poverty.' Livelihood sector report, UK: Department for International Development (DFID), viewed 12 June, 2008 <http://www.livelihoods.org/post/Docs/ABISSl3a.doc>.

Dutta, S., 2005. 'Energy as a key variable in eradicating extreme poverty and hunger: a gender and energy perspective on empirical evidence.' On MDG #1, Noida, Ghaziabad: ENERGIA International Network on Gender and Sustainable Energy.

Foley, G., 1995. 'Photovoltaic applications in the rural areas of the developing world.' World Bank Technical Paper Number 304, Energy Series. Washington DC: World Bank, p. 17.

Heltberg, R., 2004. 'Fuel switching: evidence from eight developing countries.'*...nrgy Economics*, 26(5), pp. 869–887.

IEA, 2007, *China and India insights, world energy outlook 2007*. Paris: IEA.

Joint United Nations Development Programme (UNDP)/World Bank Energy Sector Management Assistance Programme (ESMAP), 2003. 'India: access of the poor to clean household fuels', UNDP/ESMAP, Washington, DC, USA: The World Bank.

Joint United Nations Development Programme (UNDP)/World Bank Energy Sector Management Assistance Programme (ESMAP), 2002. 'Energy strategies for rural India: evidence from six states.' Washington, DC, USA: The World Bank.

Joint United Nations Development Programme (UNDP)/World Bank Energy Sector Management Assistance Programme (ESMAP), 2004. 'The impact of energy on women's lives in rural India', 2004. Washington, DC, USA: The World Bank.

Kemmler, A. and Spreng, D., 2007. 'Energy indicators for tracking sustainability in developing countries., *Energy Policy*, 35(4), pp. 2466–2480.

Leach, G., 1992. *The energy transition*. UK: Butterworth-Heinemann Ltd.

Masera, O. R., Saatkamp, B. D. and Kammen, D. M., 2000. 'From linear fuel switching to multiple cooking strategies: a critique and alternative to the energy ladder model.' *World Development*, 28(12), pp. 2083–2103.

Modi, V., 2005. *Improving electricity services in rural India: an initial assessment of recent initiatives and some recommendations*. New York: The Earth Institute at Columbia University.

National Sample Survey Organisation, 2007. *Energy sources of Indian households for cooking and lighting, 2004–05*. New Delhi: The National Sample Survey Organisation (NSSO).

Pachauri, S, 2007. *An energy analysis of household consumption: changing patterns of direct and indirect use in India*. The Netherlands: Springer.

Pachauri, S., Mueller, A., Kemmler, A. and Spreng, D., 2004. 'On measuring energy poverty in Indian households.' *World Development*, 32(12), pp. 2083–2104.

Parikh, J. K., Biswas, H. and Karmarkar, S., 2003. 'Cooking with bio-fuels: risk factors affecting health impact on rural women.' *Economic and Political Weekly*, pp. 2681–2692.

Parikh, J. K., Pandey, V. L. and Parikh, K., 2005. 'Lack of energy, water and sanitation and its impact on rural India.' In Parikh, Kirit S. and Radhakrishna, R. (eds.), *India development report 2004–05*. New Delhi: Oxford University Press.

Parikh, J. and Sharma, S., (year nm), 'Energy poverty and gender nexus: a case study of Himachal Pradesh.' India, available from www.energia.org/resources/reports/2006_wrec_irade_nexuspaper.pdf.
Parikh, J., Balakrishnan, K., Laxmi, V. and Biswas, H., 2001. 'Exposure from cooking with bio-fuels: pollution monitoring and analysis for rural Tamil Nadu.' *Energy –The International Journal*, 26, pp. 949–962.
Pitt, M., 1985. 'Equity, externalities and energy subsidies.' *Journal of Development Economics*, 17, pp. 201–217.
Ramani, K. V. and Heijndermans, E., 2003. *Energy, poverty and gender: a synthesis.* Washington DC, USA: The World Bank.
Reddy, A. K. N., 2000. 'Energy and social issues.',in *World Energy Assessment.* New York: UNDP.
Saghir, J., 2004. 'Energy and poverty.' Paper prepared for the International Energy Forum. Washington, DC: The World Bank.
Smith, K. R. and Mehta, S., 2000. 'The burden of disease from indoor air pollution in developing countries: comparison of estimates.' Paper prepared for US Agency for International Development (USAID) and World Health Organization (WHO) Global Consultation, *Health impacts of indoor air pollution and household energy in developing countries: setting the agenda for action*, 3–4 May Washington DC.
TEDDY (TERI Energy Data Directory and Yearbook), 2007. *Energy demand: domestic sector.* New Delhi: TERI.
TERI, 2003. *Rural energy matters: the Dhanawas experience.* New Delhi: TERI Press.
TERI, 2007. 'Supply of clean energy services to urban and peri-urban poor, case study' – Delhi, Global Network on Energy for Sustainable Development (GNESD), Unpublished.
The Institute of Development Studies, The University of Sussex, UK, 2003, *Energy, poverty and gender: a review of the evidence and case studies in rural China.* Washington DC, USA: The World Bank.
The World Bank, 1996. *Rural energy and development: improving energy supplies for 2 billion people.* Washington, DC, USA: The World Bank.
United Nations Development Programme (UNDP), 2007. *Human Development Report 2007/2008: fighting climate change: human solidarity in a divided world.* New York: Palgrave Macmillan.
Viswanathan, B., and Kavi Kumar, K. S., 2005. 'Cooking fuel use patterns in India: 1983–2000.' *Energy Policy* 33(8).
World Sustainable Development Forum – DSDS, 2007. *Securing energy sustainably – a development imperative.* New Delhi: TERI Press.

Part II

The global context

Trade and geopolitics

4 Trading in the world energy market[1]

Nitya Nanda

As has previously been pointed out in this book, fossil fuels are the source of most of India's commercial and primary energy supplies. They are also not a resource the country is particularly well endowed with. Thus even as 90 per cent of India's commercial energy comes from fossil fuel, domestic reserves of crude oil are sufficient for only about eight years' worth of current consumption (33 years at current production levels[2]). The situation is no different in the case of natural gas, where proven reserves stand at about 1100 billion cubic metres and would last for about 32 years at the current level of production[3]. Even with respect to coal, the most abundant energy source in India, if production continues to grow at five per cent per annum as it is doing now, proven extractable reserves are estimated to last for only about 40 years.

Meanwhile the country's consumption of non-commercial sources of energy (such as biomass) is dropping. About one-fourth of India's primary energy came from non-commercial sources in 2004. By 2031–32 however, while the total primary energy supply (TPES) will have risen substantially, non-commercial sources are expected to provide barely 10 per cent of the TPES (Planning Commission 2006).

As a result, India's dependence on foreign sources for fossil fuel (already quite high) is likely to reach about 80 per cent in the year 2031 (TERI 2006). The nation's energy security is therefore critically linked to the performance of global energy markets. The challenges in this respect are aggravated by the fact that not only has supply capacity not been able to keep up with increases in demand in the last few years (resulting in sharp price increases), but also the major suppliers of oil to India behave as a cartel.

Energy and India's trade

Before exploring the links between energy trade globally and India's security concerns, it is a good idea to understand the nature of energy flows into and out of the country. India is among the major traders in the global energy market. Of the 2235 million tons (Mt) of crude oil traded globally, India is the sixth largest importer with a share of 4.3 per cent. Yet it has very little or no influence on what is essentially a sellers' market. The country is, however, the fifth largest producer of petroleum products, with an annual production of about 108 Mt constituting about 2.9 per cent of global production (Table 4.1). In fact, India has become a significant exporter of petroleum products in recent years. The country is also not a significant trader in natural gas though it is considered to be clean and the energy of the future. This has largely been due to infrastructure constraints with liquefied natural gas (LNG) imports only commencing in 2004. Natural gas has been the fastest growing market among energy products though and currently imports stand at about 5.0Mt of LNG annually. Finally, even though India produces a large amount of coal, the size of the domestic market has meant it is also sixth largest importer of coal.

The West Asian region provides as much as 67.43 per cent of India's oil imports. It is not only the nearest region from which India can import oil, but also has about 57 per cent of the global reserves of oil, contributes about a quarter of global production, and accounts for about half of global exports. It is, therefore, rather difficult to reduce India's reliance on West Asia to any significant extent in spite of the political volatility of the region, and the inherent risks associated with concentrating supply sources.

Since India's dependence on foreign energy has been very high, energy products (particularly petroleum) have always been a significant component of India's import basket. In recent history, oil has constituted a very large share of India's total import bill. In 1980–81 as much as 40 per cent of India's import bill was due to oil imports. This share has fluctuated since and is an important index of national energy security perceptions. The share of oil in imports came down to almost 15 per cent by the late eighties before increasing once again in the wake of the first Gulf War. Post liberalization in 1991, as India's imports of other commodities grew it dropped again to about 15 per cent, primarily over the course of the next decade.

Table 4.1 India's position in global energy production and trade – 2004

	Crude oil Mt	*Coal Mt*	*Natural gas BCM*	*Petroleum products Mt*
Global production	3923	5878	28721	3719
Indian production*	34 (0.87)	430 (7.32)	32 (0.11)	108 (2.9)
Indian consumption*	130 (3.31)	454 (7.72)	32 (0.11)	107 (2.9)
Global trade*	2235 (56.97)	778 (13.24)	838 (29.18)	825 (22.18)
Indian imports**	96 (4.30)	37 (4.76)	6.9	9.75 (1.18)
Indian import dependence	73.85	8.15	–	

Source: IEA (2006).

* Figures in parentheses are percentage of global production

** Figures in parentheses are percentage of global trade.

In recent years, however, the share of oil in the import bill has once again gone up and at present stands at about 33 per cent. Interestingly, this is owing to two very different factors. The primary cause has been the sharp increase in oil prices. A second reason however, has been because of imports of crude oil meant not for domestic consumption but for refining and export.

Importantly therefore, because India is now a major exporter of refined petroleum products, the net import bill is lower. This is a relatively recent development. Even a few years back exports of petroleum products were small. In the year 2006–07, petroleum products became India's largest export item, with a share of about 15 per cent of total exports, replacing gems and jewellery, which had held the top position in the Indian export basket for decades. Between 2005–06 and 2006–07, the proportion of petroleum products in total exports has risen from about 11 to 15 per cent, while the share of petroleum in total imports has seen a marginal fall. In the fiscal year 2006–07, India's export of petroleum products was about 32 Mt, representing about 3 per cent of global trade and making India among the top ten exporters in this segment. The net oil import is less than 25 per cent of total import now. This is not very different from several countries in this regard, including the United States and Japan, though China is in a far better situation.

Having said this, it is important to point out that about one-third (35 per cent) of India's export earnings goes to finance the country's energy bill. For other major energy importers: the United States, Japan and China, the figures are 24.4 per cent, 24.8 per cent and 8.5 per cent respectively. This is because India is running a huge trade deficit. The emergence of India as an exporter of petroleum products has added a new dimension to the issue, as petroleum products constitute about 15 per cent of Indian exports, but they are entirely dependent on imported crude.

The global market and prices

Globally, oil is the most important source of energy, followed by coal and natural gas. While about 34 per cent of TPES comes from oil, coal and natural gas constitute about 25 per cent and 21 per cent of TPES respectively. The share of oil, however, has shown a substantial decline over the last couple of decades (it stood at 45 per cent in 1973). This loss of share in global consumption has been, more or less, compensated for by gains in the share of natural gas and nuclear energy, which increased from about 16 per cent to 21 per cent and less than 1 per cent to more than 6 per cent respectively (IEA 2007).

Among the three important energy products, oil is different from coal and gas, with respect to place of production and consumption. Currently about 4,000 Mt of crude oil is produced globally and about 57 per cent of that amount is traded internationally. This means more than half of the crude oil is consumed in a country other than the country of production. In the case of coal and gas, however, the situation is reversed. The shares of trade in total global production in coal and gas are about 13 and 29 per cent respectively (Table 4.1). Relatively high transportation costs, the ease of use (or rather the lack of it) or just domestic availability in consuming countries are among the reasons why this is the case. It is, however, interesting to note that though the share of trade in crude oil production is as high as 57 per cent, when it comes to refined petroleum products, the share of trade in total global production is about 22 per cent (Table 4.1), demonstrating that consuming countries prefer importing crude oil to importing refined products.

The prices of all three energy products, oil, gas and coal, were more or less stable from 1992 to 1998. In fact prices showed a declining trend during this period. They increased for two consecutive years and again showed some stability till 2003. However prices zoomed thereafter (Figure 4.1). The price of crude oil fell from $35.95 per barrel in 1980 to just $14.17 in 1986. Prices then rose substantially in the wake of the Gulf War and reached $22.99 in 1990 before falling again to as little as $13.08, the lowest price for a long period (IEA 2007). While prices did show a moderate increase till 2003, the next three years saw sharp increases, unprecedented over the last two decades.

The price of natural gas has more or less followed that of crude oil for a long time, but over the last few years it has been moderate compared to oil. In the case of coal however, the price increase has been more or less stable excepting a moderate rise in 2003 (Figure 4.1).

One important aspect of price or market behaviour of energy commodities is that they have been quite different for different commodities. As we just discussed, coal prices have moved in very different ways from those of oil and gas. In the case of natural gas it has been different across regions as well. The natural gas markets in particular have shown interesting developments. The global gas market can be segmented from two perspectives: mode of transportation and geographical region.

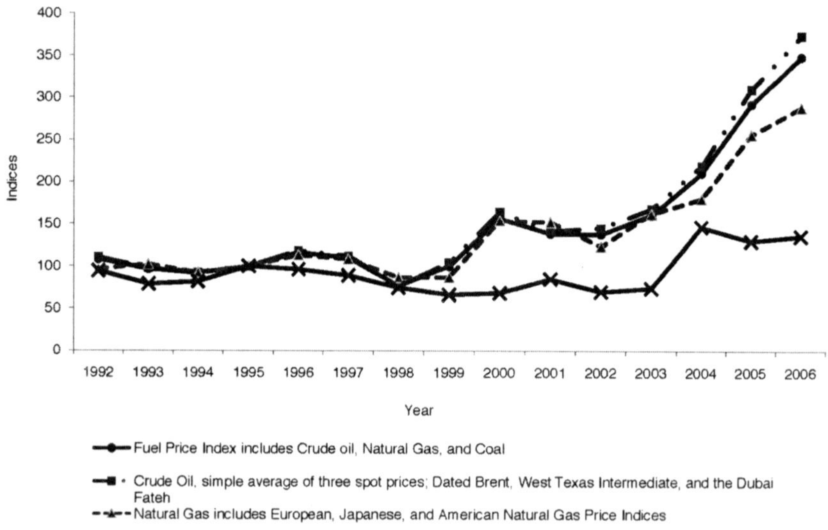

Figure 4.1 Price indices of commodity fuel, crude oil, natural gas and coal (1995=100). (Source: IMF (2006.))

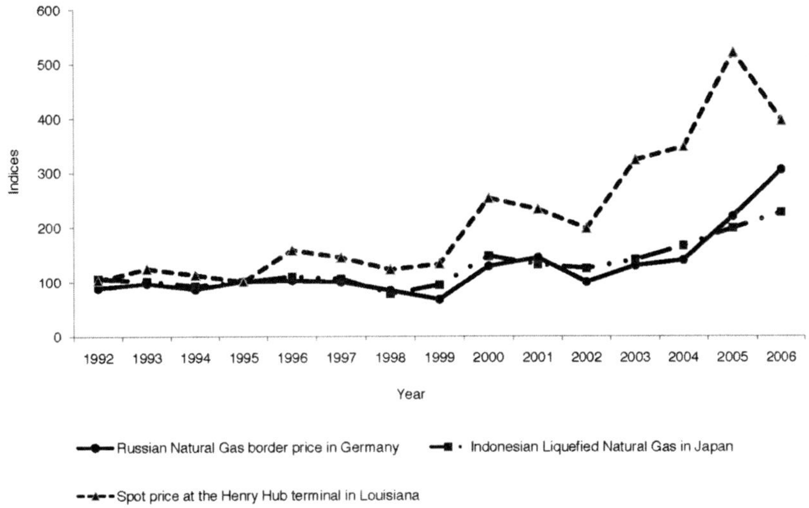

Figure 4.2 Price indices of natural gas in different markets (1995=100).
(Source: IMF (2006.))

If one compares the three main markets, those of the United States, Europe and Japan, the price increases in the United States have been sharp and unstable with wide fluctuations, while in Japan the price rise has been most the modest and least fluctuating. Europe falls between Japan and the United States (Figure 4.2). Interestingly, as far as the mode of transportation of gas is concerned, Japan is entirely dependent on LNG, while the US and European imports are overwhelmingly dominated by transportation through pipelines. This implies that prices of LNG have seen less rise and less fluctuation compared to gas transported through pipelines. Import of LNG has always been considered an expensive option compared to import of gas through pipelines. However, by the end of 2005, the price of gas imported as LNG became comparable to that imported through pipelines in the European markets and much less than the prices of gas imported through pipelines in the United States (Figure 4.2). This could be partly because the costs of liquefaction and re-gasification, which are important processes in the transportation of gas as LNG, have gone down as a proportion of the 'basic price' of gas, due to improvements in technology as well as a rise in the basic price of gas itself. It could also be due to the fact that while pipeline gas trade occurs between fixed parties, in LNG there could be options for alternative buyers and sellers, allowing some scope for competitive market mechanisms to work.

It is also interesting to note that the price of gas moved hand in hand with that of oil for some time (till about 2003), but parted with oil prices afterward. This could be because of geopolitical reasons. The importance of the West Asian region as a source of oil is much greater than for natural gas. Major exporters of gas are much more dispersed and hence a disturbance in West Asia creates much less impact on the gas price. Nevertheless, because gas is at least a partial substitute for oil, a hardening oil price has led to hardening gas prices as well. The price of coal has not been affected much as it is a poor substitute for oil and gas and also because West Asia does not export coal.

The future scenario

The moot issue in the global market for energy is whether current high prices will continue or whether they will come down.[4] It is indeed difficult to answer this question. But before one can even start answering this, it is important to answer another fundamental question: what determines the price of oil? In the 1970s, the conventional answer to this question was 'OPEC'. Similarly, since the 1980s, the answer to this question has increasingly become 'the market'. Unfortunately, it was never clear exactly how far OPEC determined prices in the seventies, nor is it clear whether only the market determines prices now. It is also a fact that OPEC continues to exist and plays a similar role even now, particularly on the supply side of the global oil market (Horsnell and Mabro 1993). Real world markets in this case do not behave in a way that some economists would like us to believe. Price formation thus involves more than just demand and supply in the market.

In the present context, several reasons have been advanced for price increases. One of them is the rising demand, particularly due to rising consumption of oil in China. Other reasons that have often been cited include perceptions about 'peak oil', and geopolitical events like the war in Iraq and violence in Nigeria, as well as the increased activities of speculators, including the role of hedge funds in the futures market (Cantrel 2006, Brodrick 2007). However, it would be interesting to look at the structure of the global energy market before looking at the other issues.

Market structure

The structure of the global markets for energy products throws some light on likely price scenarios (Table 4.2). Though market structure is normally understood from the number of firms and their relative position in the market, in terms of energy products it would be more appropriate to look at the relative position of producing or exporting countries, particularly because countries are known to use their sovereign power on production and export of energy products, especially oil. In any case, about 90 per cent of world reserves are controlled by state-owned oil companies, as opposed to market-driven public companies. Indeed, the largest independent oil company (IOC), Exxon Mobil, ranks only fourteenth on the list of proven reserve owners, behind a long list of state-owned oil companies (CFR 2006, Miller 2007).

The United States and Russia are the only two countries which are both major producers and consumers in all the three energy products. However, the basic difference between them is that Russia produces much more than it consumes and

Table 4.2 Major exporters and importers of oil, gas and coal – 2005 (in order of their share in global exports/imports)

Crude Oil		*Gas*		*Coal*	
Exporters	*Importers*	*Exporters*	*Importers*	*Exporters*	*Importers*
S. Arabia	US	Russia	US	Australia	Japan
Russia	Japan	Canada	Germany	Indonesia	Korea
Iran	China	Norway	Japan	Russia	Taiwan
Nigeria	Korea	Algeria	Italy	South Africa	UK
Norway	Germany	Netherlands	Ukraine	China	Germany
Mexico	India	Turkmenistan	France	Colombia	India
Venezuela	Italy	Indonesia	Spain	US	China
UAE	France	Malaysia	Korea	Canada	US
Kuwait	Netherlands	Qatar	Turkey	Kazakhstan	Russia
Canada	Spain	US	Netherlands	Vietnam	Italy

Source: IEA (2007).

hence is a major exporter in all the three products, but the US consumes much more than it produces, particularly in oil and gas. For example, in 2005, with a share of 7.9 per cent in global production, the United States was the third largest producer of crude oil after only Saudi Arabia and Russia, yet it was the largest importer of crude with a 25.8 per cent share in global imports (IEA 2007). China is a major producer as well as consumer of both oil and coal. India is a major producer and consumer only of coal. Overall, however, the major consuming nations and the major producing nations are, more or less, different groups of countries (Table 4.2).

Energy products are also necessary goods. Thus, they become extremely important commodities in international trade. The major buyers of energy commodities are also quite common across all three commodities (Table 4.2). There are just 13 countries that share the top ten positions across all three commodities. Comparatively, the sellers in the global energy market are more dispersed, as there are 23 countries that share the top ten positions in the three commodities. The major exporters of energy commodities, except Norway and Canada,[5] are all from the developing world, while the major importers, except China and India, are all from the developed world. The entry of China and India into the global energy trade as major buyers, however, is a recent development. While India has, all along, been highly dependent on foreign energy (oil), China became a net importer of oil only in 1993. But both became among the major buyers only around the turn of the century.

Market concentration measures generally try to capture the market power of firms by looking at indices based on market shares of individual firms. But as explained before, in the present context, it would be worthwhile to look at shares of production of individual countries in total global production as well as share of exports of individual countries in total global trade. Since all these products are finite resources, reserves are another important aspect to examine. Two measures of concentration are derived for this analysis – four-country and eight-country concentration ratios, estimated as the combined share of the top four or eight countries in production, export and reserves (see Table 4.3). It is believed that the greater the concentration, the higher will be the upward pressure on price, following the structure-conduct-performance (SCP) paradigm of industrial economics (Scherer and Ross 1990).

Currently, oil seems to be the least concentrated market compared to natural gas and coal both in terms of production and exports. When it comes to reserves, the picture however changes. While coal remains the most concentrated market, the structures of oil and natural gas become comparable. In fact, in terms of the eight-country concentration ratio, the oil market becomes more concentrated than natural gas. Iran has the second largest reserves of natural gas, yet it is not among the major (top ten) producers or exporters of gas. But it wants to enter the market in a big way. If that happens then it would definitely change market dynamics. Pressure of prices would be higher on oil than gas. Moreover, the reserves to production (R-P) ratio is lower in oil than in gas, implying that supply constraints are going to be more prominent in the oil market, putting further pressure on price.

If, on the other hand, one considers OPEC as a single country, the picture changes totally. OPEC has a share of 43 per cent in global production, 51 per cent in global trade and a high 79 per cent in global reserves. The four-country concentration ratio for oil reserves, taking OPEC as a single country, is as high as 98.5 per cent. This shows the kind of pressure one can expect on the price of oil. OPEC does not control production of natural gas and their share of global production at present is only about 18 per cent. However, OPEC countries hold about half the global

Table 4.3 Indicators of global market structure – 2005

	Oil	*Gas*	*Coal*
Reserve-Production (R-P) Ratio	44.2	64.3	180
4-Country Concentration Ratio (Production)	38.2 (67.6)	49.5 (64.3)	77.9
8-Country Concentration Ratio (Production)	55.4	61.4	92.0
4-Country Concentration Ratio (Exports)	39.9 (74.0)	54.4	63.3
8-Country Concentration Ratio (Exports)	59.2	74.5	89.4
4-Country Concentration Ratio (Reserves)	53.5 (98.5)	62.2 (82.6)	67
8-Country Concentration Ratio (Reserves)	79.7	74.4	–

Source: IEA (2007) and EIA (2006).

Note: Figures in the parentheses are concentration ratios considering OPEC as a single country.

gas reserves. If OPEC becomes active in the area of natural gas as well, its impact could be significant though less than that in the oil market.

As with current production and exports, concentration is very high in coal reserves as well, implying that the suppliers will have high market power in the long run. As of now such global market structure is not reflected in the price of coal as it is not widely traded and countries are producing it mostly for domestic consumption. It is also interesting to note that countries with high reserves or production are not the major exporters of coal. One positive aspect of the coal market is that the current R-P ratio is quite high: almost four times that of oil and three times that of gas. It may, however, be noted that a significant component of the coal reserve (21 per cent) is of a lignite type which is not easily tradable.[6] Moreover, while historically world reserves for oil and gas (more so for gas) have seen upward revisions, in the case of coal revisions have been downward.

Demand–supply gap

While, on earlier occasions of high oil prices, supply shocks were the primary reason, current high prices are largely due to high demand. So much so, that spare capacity in the production of crude oil as well as in refining has been almost eroded. According to the IEA, a 'comfortable' level of spare global capacity[7] would be five or six million barrels per day, but currently the spare capacity is less than three million barrels per day and about half of that is in Saudi Arabia (Brodrick 2007). A related factor often cited is the falling value of the dollar. Since the price of oil in the global market is denominated in dollars, depreciating dollars means cheaper oil and higher consumption for countries where the domestic currency has been appreciating vis-à-vis the dollar (Chandrasekhar and Ghosh 2008).

The growing demand for oil in China is one of the possible reasons for price increases globally, but it cannot be the only major reason. China turned to being a net importer of oil from being a net exporter in 1993. Since then it has been importing more and more oil. Yet the price of crude oil continued to fall till 1998. For the next five years the price of oil showed an upward trend, but the increase in prices was not as great as we find today. The year 2003 was the time when there was a violent break from the trend (Figure 4.1). This was also the year when the Iraq war was launched. It may also be noted that a substantial price increase was observed in the oil market even at the time of the first Gulf War, launched in 1989. But that price increase was more temporary and the price of oil in 1998 was lower even than the price level of 1988.

Peak oil

A study by Douglas-Westwood Ltd, the World Oil Supply Report, suggests that the world is drawing down its oil reserves faster than ever. At the beginning of 2003, 99 countries had produced oil or were expected to produce it in the future. Of these, 49, including the USA and Russia, are well past peak; 11, including the United Kingdom and Norway, are just beginning to see declining production; and

12, including Australia and China, will reach peak soon. The rest will see peaks within the next 25 years. In non-OPEC countries as a whole production is expected to begin declining soon. It is expected that by the end of the decade, OPEC will need to be in a position to increase its output by over 50 million tonnes per year, every year, just to offset declines in non-OPEC output and maintain current levels of production.

Even the major oil companies have been downgrading their reserve estimates. They have also failed to add any major new finds over the last decade or so. Oil companies have reduced their exploration budgets by more than a quarter. Compare this with IEA's calculations, that to meet the future demand by 2030, the required investment in oil exploration and production will be about $2.2 trillion (Fusaro and Vasey 2005).

Futures market

The existence of a futures market has generally been considered helpful as it aids in price discovery as well as in the hedging of price risk. It is also often argued that the futures market can lessen variability in prices that might otherwise be present due to seasonal fluctuations in demand and supply. The oil market, however, is not subject to much seasonality. It is also argued that the existence of the futures market may even increase volatility in the market due to the speculative market operations of the non-commercial participants in the market.

The number of non-commercial participants in the futures markets (including hedge funds), has grown alongside the volume of energy futures contracts traded. Indeed hedge funds now are more active in the oil and gas future markets than they used to be before 2003 (Fusaro and Vasey 2004). The volume of energy derivatives traded outside the traditional futures exchanges has also grown. In other commodities, such growth has been less dramatic. The average daily contract volume for crude oil as well as natural gas increased by 90 per cent between 2001 and 2006. There has also been a significant increase in the volume of energy derivatives traded outside exchanges, but this is difficult to quantify due to non-availability of information. According to the Bank for International Settlements, the notional amounts outstanding of OTC commodity derivatives excluding precious metals, such as gold, grew by over 850 per cent (GAO 2007).

Finally, the role of speculators in increasing prices in a market has received particular focus over the last few years. Some observers believe that large purchases of oil futures contracts by speculators have led to higher prices. There are others who believe that speculative trading cannot have a significant impact on prices in the long run, though it can contribute to short-term price movements. This, however, assumes that there is enough competition in the physical market. In the absence of competition, firms rarely lower prices, even when production costs decrease or demand falls. They are more likely to hold prices constant, while cutting production, than to lower them. Therefore, prices are sometimes observed to be 'sticky downward'. Indeed there is hardly any reason to believe that the oil market is competitive and there are several theoretical as well as empirical studies that indicate that oil prices may be sticky downward (Davis and Hamilton 2003). In such a situation, speculators' activities influence prices to move upwards disproportionately more than they influence downward movements. This may well be the reason that the price rise during the second Gulf War has been much sharper than the price rise in the first Gulf War. It should also be noted that speculators need a reason to go bullish. The Iraq war could have provided that reason. Moreover, continued violence in Iraq and uncertainty over Iran may have convinced speculators to hold on, leading to persistently rising prices.

Conclusion

In view of these considerations, while oil prices may soften if there is an improvement in the geopolitical situation, it is unlikely that such a shift will be marked. Even though the instability in Iraq is unlikely to have too strong an impact on oil prices as time goes on, the continued uncertainty over Iran is sufficient reason to convince speculators to remain active in the oil and gas futures market. Also, even if the Iran tension gets diffused soon, it is unlikely that oil prices will get back to 2002 levels. It is more likely that prices will gravitate towards the trend that was set in 1998. In other words, the price of oil is likely to show an upward trend. The economic recession in the United States in recent months has led to some slackening in the oil market. Nevertheless, it is unlikely that there could be a drastic fall in prices, unless of course, the global economy goes into a severe recession.

It is also now impossible for India to get cheap oil and gas. What China could get from Angola and Australia, and India could get from Qatar, are things of the past.[8] India's energy security vitally needs a smooth flow of energy products from abroad. Historically, a physical disruption of supply has never occurred. But financing a huge energy import bill is an issue. It may be noted in this context that the economic crisis of the early 1990s in India was precipitated by its inability to finance such an import bill as there was not enough foreign exchange in the country. Thus overall export performance is also a crucial element in India's energy security. A recent study has shown India to be the third most oil-vulnerable country – and that is primarily due to India's lower ability to pay (Gupta 2008). Currently, India's merchandise trade deficit is quite high and growing – hovering around seven per cent of GDP. This does not augur well for the country. Thus, India needs to improve its export performance.

Participation of Indian energy companies in the futures market can, to some extent, help these companies to hedge price risks. But much should not be expected from this, not just because taking part in the futures market itself is not free from risks, but also because futures prices prevailing now are at high levels. Promotion of a futures market for energy products in India is also unlikely to help much as prices will, by and large, be determined by what happens at other major global futures markets rather than in India. However, stronger linkages with the energy supplying nations could provide an incremental boost to exports, apart from a greater assurance of supply.

Optimising the energy-use mix in the country on the basis of future scenarios in different energy products in the global market is therefore a policy imperative. India currently has high dependence on oil and coal but very little use of gas. Obtaining oil has always been a problem for India, but much greater imports of coal in the future will also have their own problems. Given this India must explore the option of increasing the use of gas. Apart from bringing greater balance to the energy-use mix, it could also prove to be prudent as in the long run importing natural gas could be a better option than importing oil or coal.

Since India's room to manoeuvre in the global energy market is quite small, the country must try hard to reduce its dependence on foreign energy. Harnessing the potential of renewable energy not only in its own territory but also its neighbouring countries is a good strategy, not just for India but the South Asian region as a whole (Nanda and Goswami 2008). And of course, the importance of improving energy efficiency can hardly be overemphasised in this context.

Notes

1 The paper draws heavily from the work done under the project titled 'Building an Energy Secure Future for India through a Multistakeholder Dialogue Process' supported by the Nand and Jeet Khemka Foundation.
2 This includes a tentative estimate of newly found reserves at Barmer in Rajasthan reported to be about 2.5 billion barrels or about 350 Mt.
3 Geologists have predicted large amounts of undiscovered oil and gas in India, a view that has been reinforced after the discovery of oil in Rajasthan and the discovery of gas in Krishna-Godavari basin. Nevertheless, as of now, Indian reserves in both oil and gas are less than one per cent of global reserves.
4 It may be noted in this context that the real price of oil at present is still lower than its 1986 level.
5 In gas, the United States is the tenth largest exporter, but this is because of its special arrangement with Canada, while in coal, both Australia and the United States are among the ten largest exporters. Trade in coal, however, is much less significant than trade in oil and gas.
6 Now, with the availability of technology, even lignite is tradable to some extent.
7 According to IEA, spare petroleum capacity is the capacity that can be turned on for 30 days and sustained for 90 days.
8 Angola committed to supply China with 200,000 barrels per day of crude at $60/barrel for 10 years. It also secured a $25 billion contract from Australia in 2002 for supply of LNG for 25 years at a fixed price. India has been receiving gas from Qatar at a price of $3.88/m BTU, which is quite cheap.

References

Brodrick, S., 2007. 'Three triggers for higher crude oil prices.' *The Market Oracle*, 13 Jun, 2007 viewed 5 April, 2008 <http://www.marketoracle.co.uk/Article1254.html>.
Cantrel, A., 2006. 'The blame game: hedge funds and oil.' Viewed 5 April 2008, <money.cnn.com/2006/04/26/markets/hedge_oil/>.
CFR, 2006. 'Report of the Independent Task Force on the National Security Consequences of US oil dependency.' Sponsored by the Council on Foreign Relations. Washington DC: Council on Foreign Relations.
Chandrasekhar, C. P. and Ghosh, J., 2008. 'Oil prices and the dollar.' *The Hindu Business Line*, 26 February.

Davis, M. C. and Hamilton, J. D., 2003. 'Why are prices sticky? The dynamics of wholesale gasoline prices.' NBER Working Paper 9741. Cambridge, MA: National Bureau of Economic Research.

EIA, 2006. *International Energy Outlook 2006*, Washington DC: US Energy Information Administration.

Fusaro, P. C. and Vasey, G. M., 2004. 'Energy hedge funds: why have they appeared now?' *Commodities Now*, December 2004, viewed 5 April, 2008 <energyhedgefunds.com/ehfc/modules/articles-4/content/Hedge_Funds_CN.pdf>.

Fusaro, P. C. and Vasey, G. M., 2005. 'Hedge funds change energy trading.' Occasional Paper 39. Boulder: International Research Center for Energy and Economic Development.

GAO (US Government Accountability Office), 2007. 'Energy derivatives: preliminary views on energy derivatives trading and CFTC oversight.' GAO Report Number: GAO-07-1095T.

Gupta, E., 2008. 'Oil vulnerability index of oil-importing countries.' *Energy Policy*, 36(3), pp. 1195–1211.

Horsnell, P. and Mabro, R., 1993. *Oil markets and prices*. Oxford: Oxford University Press.

IEA, 2007. *Key energy statistics 2007*. Paris: International Energy Agency.

IMF, 2006. *World Economic Outlook 2006*. Washington DC: International Monetary Fund.

Miller, D., 2007. 'Will "private" oil sink the NYMEX?' Investment U Research Report. Baltimore: Investment U Research.

Nanda, N. and Goswami, A., 2008. 'Energy cooperation in South Asia: Prospects and Challenges', Man and Development, 30(1) 109–120.

Planning Commission, 2006. *Integrated energy policy*: Report of the Expert Committee, New Delhi, Government of India, Planning Commission.

Scherer, F. M. and Ross, D., 1990. *Industrial market structure and economic performance*. Boston: Houghton Mifflin Company.

TERI, 2006. *National energy map for India – technology vision 2030*. New Delhi: The Energy and Resources Institute and TERI Press.

5 Geopolitics of West Asian and Central Asian oil and gas

Implications for India's energy security[1]

Talmiz Ahmad

Energy security

A one-line definition of 'energy security' might be: an assured access to energy resources, at affordable prices, to obtain sustainable economic growth rates and economic development.

Most of the world's energy comes from hydrocarbons (oil and gas), which account for 65 per cent of the world's energy requirements. While oil accounts for 42 per cent of the global energy mix, the other sources of global energy are coal (24 per cent), natural gas (22 per cent), nuclear energy (6 per cent) and renewable and non-conventional sources (7 per cent). World energy demand has increased by 95 per cent over the last 30 years and is expected to rise by 60 per cent over the next 20 years. During this period, the demand for oil will increase by 42 per cent, while the demand for gas will increase by 97 per cent.

In recent years the most significant development in the consumption of hydrocarbon fuels is the increase in Asian demand. The anticipated rise in Asian demand in comparison with other regions is illustrated in Table 5.1, depicting the oil scenario in 2025.

Table 5.1 Global oil demand forecast: 2025 [in million tonnes per year]

	2001	*2025*
Asia: Big (4)**	946.2	1842.6
Rest of Asia	298.8	547.8
Middle East	249	448.2
North America	1195.2	1743
West Europe	697.2	697.2
Rest of world	448.2	647.4
TOTAL	3834.6	6025.8

Source: McKinsey 2005: 83.

**China, Japan, RoK and India

Note: Estimates of oil and gas reserves and projections of future demand and supply vary considerably between different source materials.

The significant change in global hydrocarbon consumption patterns is that, as against earlier years when the bulk of Asian production was consumed in West Europe and North America, today Asian countries are increasingly emerging as the principal consumers of Asian production – 55 per cent of Asian crude is being currently consumed in Asia and 95 per cent of Asian gas is being currently consumed in Asia (Mckinsey 2005: 82).

Along with oil, natural gas will play an important role in Asia's energy demand. Natural gas, being a 'clean' fuel (compared to coal), is increasingly seen as the fuel of the 21st century. At present, Asia has a much lower share of gas demand than the world average, (6 per cent versus 12 per cent). To meet Asia's rapidly increasing energy requirements, demand for gas will increase significantly (Table 5.2).

Meanwhile, the bulk of Asian demand in 2025 will be met by imports, as brought out in Table 5.3.

This chapter highlights the importance of West and Central Asia in the context of global demand for oil and natural gas. These regions are vital partners in India's growth and thus the energy scenario in West and Central Asia is key to appreciating

Table 5.2 Asian gas demand (in billions of cubic meters (bcm))

Country	*2006*	*2015*	*2025*
ASIA: Total	336	504	1036
China	56	70	280
Japan	84	112	140
RoK	28	84	196
India	42	56	84
Pakistan	28	28	56
Taiwan	28	28	70
Thailand	42	56	112
Turkey	28	56	98

Source: Ernst & Young 2006: 94.

Table 5.3 Global oil import forecasts: 2025 (in million tonnes per year)

Region/Country (% oil imported)	*Total imports*	*Net increase in imports*
Asia (74%)	1444.2	597.6
China (71%)	547.8	298.8
India (87%)	199.2	134.46
Japan (100%)	149.4	–84.66
RoK (100%)	124.5	29.88
Europe (80%)	622.5	119.52
USA (72%)	871.5	199.2

Source: PEL 2006: 84.

some of India's energy security concerns. This chapter will also discuss some of the strategies India has adopted to enhance its energy security, including importantly, a multifaceted and robust energy diplomacy. More fundamentally, we underline the underlying philosophy and perceptions of India's energy security and its inherently cooperative approach.

As part of this we focus on the core attributes of India's engagement with both West and Central Asia. This engagement has occurred against the backdrop of India's enduring interests in West Asia and is expanding in political and economic terms. We also highlight the nation's larger geopolitical interests in Central Asia. Beyond the West and Central Asian region, we look at India's efforts at the multilateral level to address its energy security, at forums such as the Shanghai Cooperation Organization, the Asian Producer-Consumer Dialogue and so on.

We conclude by highlighting some considerations India must keep in mind while striving to enhance its energy security vis-à-vis West and Central Asia, such as the political instability in Iraq, the Caspian Sea's legal status and the need to engage other players present in the energy sector in the two regions.

West and Central Asia in the global energy supply scenario

West Asia

The Gulf provides the bulk of the world's oil: just five countries of the Gulf (Saudi Arabia, Iraq, Iran, Kuwait and the UAE) have about 65 per cent of the world's reserves. The Gulf has proven reserves of 92 billion tonnes of oil, and production capacity of 1110.5 million tonnes a year; uniquely, it also has over 90 per cent of the world's excess production capacity.[2]

Imports from West Asia have been significant in the global oil trade (Table 5.4). Production costs in Gulf OPEC nations are less than $3 per barrel (one metric tonne is equivalent to about 7.3 barrels and thus this cost works out to about $22 per tonne), and the capital investment required to increase production capacity by one barrel per day is less than $5,940. This may be compared with a capital investment of $10,000–$12,000 needed to produce one barrel per day outside the Gulf region (Janabi 2004). Thus Gulf OPEC producers can expand capacity at a cost that is a relatively small percentage of projected gross revenues. According to the International Energy

Table 5.4 Oil imports from the Gulf (%)

Year	*USA*		*West Europe*		*Japan*	
	% of total demand	*% of total* imports*	*% of total demand*	*% of total* imports*	*% of total demand*	*% of total* imports*
1990	11.3%	28%	29%	45%	66%	65%
1995	9%	20%	23%	44%	79%	70%
2000	12.6%	24%	21%	42%	75%	75%

Source: Fesharaki 2005: 101.

Agency (IEA), the Gulf region's oil production capacity, between 2005 and 2030, would expand from 27 MBD to 45 MBD (1,344 to 2,241 million tonnes per year). The region would then represent 40 per cent of the world's oil supply, and would account for three-fourths of additional global supply from 2000–2030 (ECSSR 2007).

The principal sources of global gas also lie in Asia. The Asian area of Russia (North Asia) has 27 per cent of the world's proven reserves, followed by Iran (15 per cent) and Qatar (14 per cent). In fact, North and Central Asia and the Gulf between them have over 70 per cent of world reserves (Table 5.5).

Central Asia and the Caspian Sea region

The area around the Caspian Sea has shown evidence of oil resources for at least two millennia. The Caspian region is abundant in oil resources and as per EIA data, the total proven oil reserves in the Caspian Sea region ranges from 2,320 to 6,683 million tonnes and total proven natural gas reserves are estimated at 6,496 billion cubic meters (EIA 2007). This amounts to four per cent of the total oil reserves of the world and five per cent of the total gas reserves.

While there is considerable international focus on Central Asian hydrocarbon resources, it is important to recognize their limitations both in terms of quantity and accessibility. This is illustrated in Table 5.6, comparing Gulf and Central Asian reserves and projected production.

While the Caspian area is not at the level of West Asia as a source of supply, it is important to the energy policy of many nations because of its position as a significant and alternative source of oil and gas supply to the West Asian region.

India's energy scenario

Hydrocarbon Vision 2025, published by the Government of India in February 2000 (GOI 2001), set out in stark terms India's energy security predicament: its crude oil self-sufficiency had declined from 63 per cent in 1989–90 to 30 per cent

Table 5.5 Global gas reserves (in trillions of cubic metres)

Region	*tcm*
Russia/Central Asia	53.2
West Asia	56
Africa	11.2
South & South East Asia	8.4
North America	5.6
South/Central America	5.6
Europe	4.2
Australia	2.8

Source: Ernst & Young, 2006: 94.

Table 5.6 Gulf and Central Asian oil and gas reserves

			Gulf	*Central Asia*
(A) Oil	(a)	proven reserves	92350 Mt	2182 Mt
	(b)	% of global reserves	65%	1.5%
	(c)	undiscovered reserves	27825 Mt	4678 Mt
	(d)	% of global reserves	28.2%	4.7%
(B) Gas	(a)	proven reserves	52 tcm	7.42 tcm
	(b)	% of global reserves	34.4%	4.9%
	(c)	undiscovered resources	35 tcm	10.2 tcm
	(d)	% of global resources	23%	7%

Source: Abi-Aad 2002.

in 2000–01. The situation is likely to get worse in the future and is equally grim in the case of natural gas.

The *Integrated Energy Policy* document published by the Planning Commission in August 2006 takes a holistic view of India's energy requirements needed to meet a minimum growth rate of eight per cent per annum until 2031–32, i.e., the end of the 15th Five Year Plan. The Planning Commission has pointed out that, given the country's commitment to eradicating poverty and empowering its people with education and health, rapid growth rates of around eight per cent per annum over the next 25 years are essential to attain these goals.

In order to fuel a sustained annual growth rate of over eight per cent, India would need to ensure that basic capacities in the energy sector and related physical infrastructure, such as rail, ports, roads and water, grow by factors of three to seven times by 2031–32, alongside a 20-fold increase in nuclear and a 40-fold increase in renewable energy (Planning Commission 2006). In all scenarios, *fossil fuels* will be between 74 per cent and 85 per cent of India's energy mix, as against 96 per cent at present.

Taking into account the energy requirements detailed by the Planning Commission, efforts at enhancing India's energy security can be outlined as follows:

- augmenting domestic resources
- maximizing the use of the national hydropower potential
- obtaining the materials and technology to pursue civilian nuclear power projects
- pursuing energy efficiency and demand side management policies, and
- diversifying energy sources through increased use of renewables.

Given the high level of dependence on imports and the need to obtain internationally developed technologies to enhance the country's domestic resources and capabilities, India has committed itself to pursuing a robust 'Energy Diplomacy'. This consists of substantial, pro-active and multifaceted engagements across the world to promote India's energy security interests. The overseas engagements are aimed at achieving the following.

- Significant enhancement of domestic resources and capabilities by bringing in state-of-the-art foreign technology and expanding the national knowledge base.
- Acquisition of assets abroad, involving either equity participation in producing fields; and/or exploration and production (E&P) contracts in different parts of the world, both onshore and offshore.
- Participation in downstream projects (refineries and petrochemicals) in producer and consumer countries on the basis of criss-cross investments.
- Finalization of long-term LNG contracts.
- Setting up of transnational gas pipelines.
- Obtaining technologies to promote sustainable energy use, including conservation, increased use of environment-friendly fuels, and development of unconventional and non-conventional energy resources within the country.
- Promotion of intra-Asian dialogue between producers and consumers, encouragement to intra-Asian investment, and development of Asian capabilities, resources and infrastructure.

India's perceptions of energy security

India does not see the pursuit of its own or the Asian and global energy security interests in competitive terms and in terms of conflict. It is true that given the central role that energy security plays in the national development of a country, it has to be seen as an integral part of the national security of the country concerned.[3] However, while national security has at its core the maintenance of a country's *national* interest, energy security cannot be attained on a purely national basis; it is inherently *cooperative* in character and is founded on engagements with other countries, especially given that hydrocarbon resources will continue to dominate the global energy mix (and, hence, the energy mix of most countries) for at least the next 25 years, if not longer. For the world's energy resources to be harnessed efficiently, a cooperative approach at bilateral, regional and international levels is both inevitable and urgent.

India's commitment to a cooperative approach in the pursuit of energy security interests is strengthened by the realization that the next few years will see a steady decline in oil supplies, with consequent implications for prices, economic programmes and increased political contentions. Though advances in technology will provide the hydrocarbon resources required to meet global demand, at least over the next 30–50 years, new oil will be available in physically challenging areas such as the deep sea or frozen terrain or environmentally sensitive locations. Again, this will require huge investments for its extraction, amounting cumulatively to about $5 trillion up to 2030, at the rate of $20 billion per annum. Meeting the global demand for oil and obtaining the financial resources to ensure supplies requires the rejection of political contentions based on narrow national considerations and, in their place, calls for an integrated regional and global effort to pool together the world's human, financial and technological resources in a spirit of cooperation for mutual benefit.[4]

In line with these perceptions, India's energy diplomacy seeks to pursue engagements at bilateral, regional and global levels to promote corporate joint ventures and government-to-government partnership to encourage dialogue and the realization of mutually beneficial outcomes. At the same time, India remains conscious of the strategic dimensions arising out of the pursuit of energy security interests by different stakeholders, and, through its bilateral and regional engagements, has been attempting to ensure that its broad national interests are safeguarded.

The Gulf and Central Asia have a crucial place in the pursuit of India's energy security interests. Both these regions attract considerable international attention on account of their significant hydrocarbon reserves. It is therefore not surprising that both regions are politically volatile in terms of their internal situation and as a result of the interplay of external forces that are competing in these regions for power and influence. India's long-term interests lie in setting up alliances and partnerships, particularly in the Gulf and Central Asia, that would bring together different capabilities through joint proposals. This cooperation ranges across the hydrocarbon value chain, and includes prospecting in each other's territories, and exchanges in regard to R&D, technology, safety norms and training. Beyond the bilateral aspect, it includes the possibility of Indian and foreign national companies working together on specific projects in third countries.

India's energy diplomacy in the Gulf and Central Asia

The Gulf region provides two-thirds of India's oil requirements. From amongst these countries, Saudi Arabia is India's largest supplier of crude oil, meeting 25 per cent of its annual requirements. Following the visit to India of King Abdullah bin Abdul Aziz in January 2006, the two countries have agreed to transform their present commercial ties into a 'strategic energy partnership' through investments in each others' downstream and petrochemicals projects, as also through India's participation in Saudi Arabia's upstream proposals in the gas sector.[5] Besides Saudi Arabia, Iran is India's other major energy partner in the Gulf; it is the third largest supplier of oil to India (at 2.5 Mt/year, after Saudi Arabia and Nigeria), and is emerging as an associate in a number of hydrocarbon related projects, particularly in the gas sector. Other major partners of India are the U.A.E., Kuwait and Oman, with Qatar emerging as an important source of gas.

In an effort to diversify its energy partners, India has in recent years turned its attention to Central Asia and Turkey. In 2005, the Indian Petroleum Minister made highly publicized visits to Kazakhstan, Azerbaijan and Turkey, which led to the finalization of a comprehensive programme of bilateral and regional cooperation in the hydrocarbon sector. In Kazakhstan, Indian companies are pursuing E&P proposals in the Caspian, while Azerbaijan and Turkey are emerging as partners in the gas pipeline sector in the execution of pipeline projects from Central Asia to Europe. Indian and Turkish companies will also cooperate in executing projects in their own countries and in third countries. The possibility of purchasing Azeri oil at Ceyhan was also examined: this led to the first shipment of Central Asian oil to India, in late 2005.

India and the Gulf

India has enduring interests in the Gulf and the Arab Peninsula. The region is part of India's *security perimeter* and developments in the region have a direct bearing on India's strategic and security interests. Additionally, the Gulf and Arab Peninsula are the principal source of India's *crude oil* requirements, providing over two-thirds of annual imports. Thus, the security of oil facilities and of the sea lanes is a crucial element in India's long term *energy security* interests.

The region is also a major *economic partner* and a market for Indian goods, a partner in joint ventures and technology transfer arrangements, and a principal source of remittances from the resident Indian community. Between 2001 and 2006, India's total non-oil trade with the GCC countries rose three-fold – from US$5.55 billion in 2000–01 to US$16.35 billion in 2004–05, due to buoyancy in both exports and imports (Exim Bank 2006). In 2006–07, India-GCC two-way non-oil trade was $47.4 billion. Concerns relating to the welfare of India's four-and-a-half-million strong *Indian community* also require that India maintains the closest possible political ties with the countries of the region.

From the late 1990s, there has been an increasing recognition in the Gulf of India's political and economic strengths based on the consolidation of the country's democratic and multicultural order, its high growth rates and its firm technological base, all of which have helped convince the GCC countries that India was poised to play an enhanced regional and even international role in the economic and technological arena. These perceptions led the GCC Foreign Ministers to invite India as a dialogue partner, only the fourth after the United States, European Union and Japan. Following this, the first GCC-India Industrial Conference was held in February 2004, when six Commerce and Industry Ministers from the GCC countries, and the then Indian Minister for Commerce and Industry, agreed, through the 'Mumbai Declaration' to enhance economic cooperation by focusing on four select priority areas: trade; investments; industrial cooperation including small and medium enterprises (SMEs)/small scale industries (SSI); and transfer of technology including information technology.

India has also engaged with specific Gulf countries in high-value joint projects such as the billion-dollar Oman-India Fertilizer Project and the agreement with Qatar to purchase LNG over the next 25 years. Besides these mega-contracts, Indian companies have significantly expanded their economic presence in the region not only through enhanced trade activity but also through several joint ventures in industrial production, consultancy, information technology, engineering, management and accountancy services.

These expanded economic ties have moved along with enhanced political links. Commencing with the visit of the then Indian External Affairs Minister, Jaswant

Singh, to Saudi Arabia in January 2001, there has been a steady expansion in high-level political and economic dialogue. Between 2005–08, the Heads of State/Government of every one of the six GCC countries visited India. The sinews of the relationship are being strengthened with regular meetings of the bilateral Joint Commissions and Foreign Office Consultations, along with discussions in regard to agreements in the economic, civil aviation, health, education, security and criminal areas.

India and Central Asia

The former Indian Foreign Secretary and National Security Adviser, J N Dixit, in November 2003, succinctly set out what should constitute the main elements of India's relations with the Central Asian countries (Dixit 2004).

- India has to undertake an attitudinal change towards Central Asia. It must, as it has focused attention on the Association of Southeast Asian Nations (ASEAN) countries, perceive the Central Asian countries as an *integral part of its neighbourhood.*
- India's policies towards Central Asia should be predicated on the perception that multifaceted relations with Central Asia are essential for the security and stability of the integrated strategic region extending from the northern limits of the Central Asian Republics to the southern extremities of the South Asian region.
- India must be more active in associating with new economic cooperation and regional security arrangements which are being initiated by the Central Asian Republics.
- India must actively pursue the establishment of economic, technological and defence-supplies cooperation with these countries.
- These efforts of India must be backed up by substantive programmes and projects of educational, cultural, and scientific and technological cooperation.
- India should enhance air connections with Kazakhstan and Uzbekistan.

Since the emergence of the Karzai-led government in Kabul, India's focus on Central Asia has shifted from security to business. India's total bilateral trade with the region is now about US$300 million per annum, primarily dominated by trade ties with Kazakhstan.[6] In spite of considerable effort over the last few years, India's presence in the Central Asia energy sector remains modest. It is limited to a 15 per cent holding by ONGC Videsh Limited in Kazakhstan's Alibekmola oil field and a 10 per cent holding in the country's Kurmangazi field. Recently, the Indian company, Gas Authority of India Limited (GAIL), has been awarded two fields for exploration of gas in Uzbekistan in April 2007.

Besides commercial interests in Central Asia, India continues to have political and strategic interests, centred around curbing the influence of Islamic extremism, particularly the influence of Taliban-related activities, and contributing to the maximum extent possible to a unified stable and peaceful Afghanistan. This Indian agenda finds a ready resonance among the Central Asian countries, which has led to a brisk exchange of high level visits and expanding political and economic ties. With Iran, there has been a regular exchange of visits between Indian and Iranian leaders. Indian Foreign Minister Pranab Mukherjee was in Iran on 6–7 February 2007, while the Iranian Foreign Minister had visited India in November 2006. Foreign Minister Mukherjee not only highlighted the importance of ties with Iran for India, he also described Iran as a factor for stability in the region. India has been vigorously pursuing the *Iran–Pakistan–India gas pipeline project* over the last two and a half years in spite of sustained US public criticism of the project and diplomatic demarches. In fact, this project could possibly include Russia and China as active participants: senior Russian officials and corporate heads have repeatedly indicated the interest of their companies in participating in the project. Chinese commentators have also publicly stressed the need for India and China to pursue their own linkages with Iran. According to Liu Xuecheng, director of the Beijing Centre for American Studies and a senior fellow of the China Institute for International Studies (quoted in Varadarajan 2006):

> The US is trying to coordinate with all countries around Iran in order to isolate it … . India and China have good relations with the US but must follow their own strategy on Iran. If India gives up on Iran in the hope of securing nuclear energy from the US, it may end up with nothing. You would lose your strategic pipeline and the US might also abandon the nuclear deal at some point in the future. Pipelines from Iran and Central Asia are a strategic lifeline for Indian energy security.

India's multilateral engagement and strategies

Beyond the robust ongoing bilateral interaction based on high level visits, joint commissions and foreign office consultations and exchanges of economic delegations, India's engagement with both the Central Asian region and the larger world is taking place at the multilateral level on a number of platforms.

First among these is the Shanghai Cooperation Organization (SCO), where India has 'Observer' status. The SCO clearly provides a valuable institutionalized arrangement for Russia and China to ensure that their interests, in a region of considerable strategic and economic importance to them, are not abridged by expanding American influence. On their part, the Central Asian countries, too, see the Russian and Chinese presence as providing a valuable counterweight to the American influence, thus guaranteeing them much more autonomy. To quote from Roy 2006, the SCO essentially provides 'a delicate equilibrium among the members in the post-Cold War geopolitical paradigm'.

India has also been participating in the Trilateral India–Russia–China Dialogue on a regular basis since September 2002. Despite being a low-key affair, the regularity of the dialogue over the last five years at the Foreign Ministers' level (and once at summit level) confirms that the three countries do attach considerable importance to it as a platform for a candid exchange of views on matters of common interest. The leaders of China, Russia and India had their first Trilateral Summit in July 2006 in St. Petersburg, on the margins of the G-8 Summit. In a briefing on the meeting, the Chinese Assistant Foreign Minister said that the three leaders had met at the Trilateral Summit because they saw it as 'beneficial to boosting the cooperation among the three countries as well as maintaining multi-polarity, peace, security and development in the world.'[7]

Regional energy-related projects are another forum for greater cooperation within the region. India attaches great importance to gas pipelines from Central Asia. India is conscious of the fact that *Turkmenistan–Afghanistan–Pakistan–India* (TAPI) is an important project not just in terms of meeting Pakistani and Indian energy needs but also because it would provide Afghanistan with transit fees and possibly even the chance to develop Afghanistan's own gas reserves in the Herat region. The project has considerable geopolitical significance in that, for the first time, South Asia would have access to gas from Central Asia. Once the pipeline is operational, it is possible that Turkmenistan could evolve from a single source of gas to the pipeline into a regional hub, with pipelines from neighboring countries, such as Uzbekistan, Kazakhstan, Azerbaijan and even Russia, linking up with the pipeline to meet the increasing demands of South Asia. In due course, pipelines from the Caspian could also go to LNG terminals on the Gulf to transport Central Asian LNG to South East Asia and North East Asia.

The Iran–Pakistan–India (IPI) gas pipeline project has a sound commercial base as Iran has the world's second largest gas reserves, particularly offshore in the South Pars and North Pars fields (which it shares with Qatar). A pipeline from the Iranian collection center of Assaluyeh on the Gulf to the Indian border would be about 1,900 km, which is well within the range of economical gas supply by pipeline vis-à-vis LNG. The main point to be noted is that the IPI project is not a trilateral Government-to-Government project; it would be owned by an international consortium made up of the national companies of the three countries concerned together with other international partners, possibly companies from China and Russia and the international mining giant, BHP-Billiton of Australia. The project would operate within the framework of international commercial law, reinforced by the trilateral 'Framework Agreement' and the three governments' commitment to the provisions of the Energy Charter Treaty. The ownership of the project by an international consortium and the fact that Pakistan itself would be a major consumer of the gas would severely limit Pakistan's ability to tamper with the pipeline.

Another key forum for energy diplomacy is the Asian Producer-Consumer Dialogue. India took the first significant step in setting up a platform for dialogue between the principal Asian oil and gas producers and consumers when it convened a Round Table of Asian Oil Ministers in New Delhi, in January 2005. In his inaugural address, the then Indian Minister of Petroleum and Natural Gas, Mr. Mani Shankar Aiyar, noted that Asian producers and consumers would obtain stability, security and sustainability in the Asian oil and gas economy through mutual interdependence, which called for 'mutual investments in each other's countries by producer and consumer nations of the Asian oil community'. The Minister envisaged a future 'in which Asian countries can become major participants in the massive investments which need to be made in the Asian countries represented at this table, running to an estimated 1,580 billion dollars over the next 25 years in upstream, midstream and downstream oil and gas development in all our countries put together' (Aiyar 2005). Some of the agreements at the Round Table were the following:

- Asian cooperation in the oil and gas economy must include 'moderation, dialogue, mutual understanding and respect, security of international supplies, demand-supply equilibrium, and strategic partnerships based on a reciprocity of interests'.
- Prices should be sustained at levels which encourage Asian consumers to increase their purchases of Asian produce; at the same time, prices should be such as to encourage Asian producers to promote investment in oil and gas for Asian consumer destinations as an economic priority.
- Crisscross investments all along the entire oil and gas products chain through reciprocal investment interlocking of producers and consumers will guarantee security of both supply and demand, thus contributing to stability of prices and thereby security of both supply and demand.
- The Round Table endorsed the importance of energy conservation for the protection of the environment and issues of climate change, and the need for technological cooperation in the pursuit of cleaner and more environmentally sound fossil fuel technologies.

A Second Round Table was held in New Delhi, in November 2005, when the representatives of the same four principal Asian consumer countries met Ministers from North Asia (Russia) and Central Asia. The assembled dignitaries agreed to study 'the promotion of developing gas and oil interconnections through LNG and through transnational oil and gas pipelines within the Asian region for integration of energy markets as well as improving the transportation infrastructure'.

While the world's gas map depicts numerous pipelines moving across thousands of kilometres from Russia, Central Asia and the North Sea to Western Europe, hardly any pipelines move eastwards and southwards. The *Asian Gas Grid* envisages the setting up of a series of pipelines that will carry natural gas from North and Central Asia and the Gulf to various consumption centres in East and South Asia. Some of the principal pipelines in the proposed grid are the Russia–Kazakhstan–Iran pipeline to the Gulf, the Myanmar–China pipeline, the Sakhalin–Nakhodka–ROK–China pipeline, the Uzbekistan–Turkmenistan–Azerbaijan pipeline, the extension of the BTC pipeline from Ceyhan through Syria and Jordan to Egypt, the Iran–Pakistan–India pipeline, and the Turkmenistan–Afghanistan–Pakistan–India pipeline (Ernst & Young 2006: 90–102).

According to current estimates, the additional pipelines required to realize the Asian gas grid would be about 22,500 km, costing about $22 billion. The Asian continent, particularly Russia, the principal Asian consuming countries, and the major producing countries of the Gulf are readily able to provide the financial and technological resources for this project. Besides contributing significantly to the growth and prosperity of the continent as a whole, the project would have other benefits such as developing the electricity, petrochemical and fertilizer industries; promoting trans-continental cooperation in trade and industry; upgrading local skills and expanding employment opportunities; and, above all, generating the resources that would enable Asian governments to fund their poverty alleviation and other welfare programmes.

In simple financial terms, the total financial benefits in the first year of the project (2006) would be $1.5 billion. While the economic advantages of the gas grid are obvious, it has to be accepted that transnational pipelines are difficult and complex ventures, which are particularly daunting in an Asian environment that has been the theatre of considerable intracontinental discord and conflict, and has relatively few success stories in regard to regional and continental cooperation. At the same time, it should be noted that the international community, over the last 35 years, during which thousand of kilometres of gas pipelines have been laid across all our continents, has developed laws, rules, norms and practices that ensure that pipelines can be insulated to a considerable extent from the vagaries of day-to-day politics and made 'safe and secure'on the basis of international best practice.

Finally, recent efforts towards substantive reforms in the Asian oil market have also created a space for large consumers and producers of oil in Asia to work together. When the Asian oil ministers met in January 2005, Indian Petroleum Minister Aiyar pointed out that Asia was no longer a residual consumer of Asian oil production and yet Asia had little in terms of a global market, and even less in terms of a well-prepared oil and oil products market (Aiyar 2005). The assembled ministers were also conscious of the fact that the surge in oil prices over the previous year, even as it had adversely affected economies in different parts of the world, called for greater scrutiny of the organization and functioning of the world's oil markets – their non-transparent and non-rational foundations and procedures. The need for reform was expressed strongly, particularly by developing countries that were seeing their hard-earned resources wither away and their development programs in jeopardy. The Japanese Minister of Economy, Trade and Industry called for the development of oil markets in Asian countries so that pricing mechanisms would emerge that would correctly reflect the supply-demand balance in the Asian region (GOI 2005: 24). Following the Round Table, India and Japan agreed to jointly fund the preparation of a report on reforming the Asian oil market.

Enhancing India's energy security in West and Central Asia: some considerations and concerns

Viability of Central Asian energy resources

The Central Asian region poses several problems for countries interested in diversifying their energy security, as compared to the Gulf. To begin with, Central Asia lacks the technical infrastructure required to ensure rapid petroleum development. The development of oil and gas fields in Central Asia is expensive as compared to the Gulf: the fully built-up cost of Central Asian oil is $12.15 per barrel;

Gulf costs are two to three times less. In addition, Central Asian countries are land locked and primarily depend on proximate markets; and while Central Asian gas could be sent to South Asia and East Asia through pipelines, these would be fairly expensive projects.

The continued uncertainty over the Caspian Sea's legal status is also hindering further oil and gas development in the area. The five littoral states have failed to agree on a plan to divide the sea's resources, including the oil-rich seabed. Azerbaijan, together with Russia, and Kazakhstan, have advocated establishment of maritime boundaries based on an equidistant division of the sea, but Iran and Turkmenistan disagree (we draw upon APS 2006 in this discussion of the historical and legal background of the Caspian Sea). The Caspian Sea has six separate hydrocarbon basins, although most of its oil and gas reserves have not been developed as yet. The northern part of the Caspian is said to contain more hydrocarbons than the southern part.[8]

Three options to resolve the matter have been proposed. First, joint ownership of both the seabed and surface of the Caspian; second, division of the seabed and common sovereignty over sea water; and third, dividing up both the seabed and water. Legal experts in the region have argued that since the first option is not practicable, the choice must be between the second and third alternatives. Russia favours dividing up only the seabed, with common sovereignty over the sea's surface, because it will provide the best security. Much to Tehran's chagrin, Russia has divided the seabed with Kazakhstan, accepting a share of 16.5 per cent. Tehran goes along with the idea of joint sovereignty over the surface, but wants an equal division of the seabed between the five littoral states: an equal division of the seabed would give Iran a share of 20 per cent, whereas its shoreline covers only 13 per cent.

A third area of concern pertains to the viability of transnational pipelines, though as we have argued earlier, there are also good reasons to go ahead with such projects. Current concerns include the possibility of attacks on pipelines passing through the disturbed province of Baluchistan in Pakistan and concerns in India that, amidst possible deteriorating bilateral relations, the Pakistani Government could 'turn off the tap' and deny India the gas needed for its power and industrial projects. Similarly, with regards to a proposed pipeline project to India from Turkmenistan, via Afghanistan and Pakistan, there are doubts as to whether Turkmenistan has the gas reserves to justify the pipeline and whether it is legally able to make these exports in light of questions as to whether prior commitments to Russia take precedence. The uncertain situation in Afghanistan also inhibits progress on the project.

Finally, given the prolonged insurgency in Iraq after the capture of Baghdad in May 2003, many US plans relating to the setting up of a popular political order in Iraq and the utilization of Iraq's oil resources for local development (and possibly to finance projects which the US Administration may have made for the region), have not been implemented. Consequently Iraq's oil industry remains in a parlous condition.

To put the Iraqi oil industry in some order, the Iraqi Cabinet on 26 February, 2007 approved a Hydrocarbon Law in which it upheld broad federal control over the Iraqi oil industry. The proposed law envisaged that the Ministry of Oil would have an administrative, advisory and regulatory role, with executive decision-making and policy execution vested in the Federal Oil and Gas Council (FOGC). A newly created Iraqi National Oil Company (INOC) will be responsible for exploration and production in Iraq on behalf of the government.[9] According to some commentators, the proposed law seems to have a number of ambiguities in crucial areas, including revenue sharing arrangements with the provinces, particularly the Kurdish region, as also the financing of INOC and its role in Iraq's oil sector. Some critics have suggested that the proposed law is so skewed in favour of foreign firms that it could end up heightening political tensions in the country and spreading instability. For example, it specifies that up to two-thirds of Iraq's known reserves would be developed by multinationals under contracts lasting for 15–20 years (Mekay 2007).

Iraqi labour leaders have pointed out that transferring ownership to foreign companies would encourage the US to continue its occupation of Iraq on the ground that the foreign companies would need to be protected. Above all, there are widespread concerns that the law sets the Iraqi oil industry firmly on the path to full privatization.

Engaging other principal players in West Asia and Central Asia

China

China is a key player in the arena of energy geopolitics in West and Central Asia. Given China's energy security interests, it is not surprising that it is pursuing policies of robust engagement with the principal oil producing countries of the Gulf. Some American commentators have envisaged nightmare scenarios in West Asia in terms of which China would vigorously pursue alliances with US's adversaries or even dilute US presence in the region. Others have gone further to envision a protracted US–China struggle for oil, with a close China–Saudi relationship that could lay the groundwork for a world war (Chanlett-Avery 2005). West Asia expert Kenneth Pollack has also spoken of a trend in the region to turn to China as an alternative to the US. According to him, 'a lot of Arab states … are looking to China not just as a potential economic partner, but also as a potential political counter weight to the US,' (Massoud 2006). As Leverett and Bader have warned, if Sino–US competition is not managed prudently, it will 'generate multiple points of bilateral friction and damage US strategic interests in the region' (Leverett and Bader 2005).

In contrast to this strong rhetoric, it is interesting to observe that certain Chinese and even American commentators have recommended a more active Chinese role in West Asia. Thus, Prof Wu Lei, speaking at the China–Arab Cooperation Forum, in December 2005, had strongly recommended that China should enhance its economic and trade relations with the Middle East and North African nations, which presently are limited to energy, and 'vigorously get involved with the region's affairs for promotion of peace and stability' (Wu Lei 2006). Such long-term strategic energy cooperation, Lei said, 'will pave the way for both partners to further collaborate in other areas like politics, culture, science and technology. Needless to say, energy cooperation between China and the Arab nations goes well beyond energy itself. Geopolitical considerations are evident for facilitation of dialogues and coordination with each other's energy policies.'

China has begun to assert its interests in the hydrocarbon resources of Central Asia. China also shares Russia's concerns relating to ethno-nationalism and radical Islam, which could have a deleterious impact on the Islamic population in its North-West provinces and the Autonomous Territories. Geographical proximity to the region's hydrocarbon resources has made it natural for China to pursue cooperative exploration and production ventures and pipeline proposals with different Central Asian republics. In the face of American presence in the region, China and Russia are increasingly pursuing a partnership with each other. Both see in the Shanghai Cooperation Organization (SCO), the potential for a bloc which could thwart US attempts at regional hegemony.

Chinese efforts at setting up hydrocarbon-based partnerships with Central Asian countries have already met with dramatic success. In December 2005, an oil pipeline of 1240 km from Kazakhstan (Atasu) to Alashankou in North-West China was inaugurated. Again, signalling Sino–Russian partnership in the region, China has asked Russian companies to fill the pipeline with oil till Kazakhstan is able to do so. Thus, for some time, half the oil pumped through the pipeline (having a capacity of 9.96 million tonnes a year) will come from Russia. The capacity of this pipeline will later be downloaded, and it will be extended 3,000 km across Kazakhstan to the Chinese-operated oil fields near the Caspian Sea.[10] Later, China purchased the Kazakh company, Petrokazakhstan, for US$4.18 billion. Apart from this, from 1997, Chinese oil companies have invested more than US$2.6 billion in Kazakhstan.

China has also moved to access gas reserves from Turkmenistan, with a contract for supply of 30 bcm of Turkmen gas annually from 2009 through a pipeline which will go through Kazakhstan, linking up with the existing Bukhara–Tashkent–Almaty pipeline and extending it to the Chinese border at Alashankou (Olcott 2005).

Moscow's interest in Uzbekistan's gas seems to have sparked increased Chinese interest as well, with the China National Petroleum Company (CNPC) signing a $600 million agreement with Uzbekneftegaz for 23 smaller oil fields in the Bukhara area (Olcott 2005). China is also moving closer to Azerbaijan and has become a partner in the oil and gas sector. Several Chinese companies have already been granted production-sharing agreements by the Azerbaijan State Oil Company for the development of onshore oil fields in the country. For instance, in June 2004, the Chinese oil company, Shengli, received permission to work on the Garachukhur oil field. It is expected that increase of Azeri oil and gas production will further expand Azeri–Chinese relations (Cornell and Ismailzade 2005).

The United States

Moving on from China, we consider the role of the United States in this part of the world. The United States sees Central Asia as a region with considerable hydrocarbon potential that could emerge as an alternative source of oil and gas and thus support its efforts to reduce its hydrocarbon dependence on West Asia. The region is also important on account of the depletion of production in Alaska and the North Sea (Arvanitopoulos 1998). The United States has some specific reasons to be interested in Central Asian hydrocarbon reserves.

First, the oil of the region is of good quality and domestic consumption is not very high, thus making large quantities of local production available for export. In addition, since the countries of the region need capital and technology to develop their reserves, there are attractive opportunities for Western oil companies to participate in petroleum-related projects in the region (Arvanitopoulos 1998).

From the geopolitical perspective, the US interest is in consolidating its own position in the region, wrenching the region from Russian influence, ensuring that Iran is excluded from the major oil and gas development projects and transportation routes, and, above all, ensuring that the bulk of Central Asian production finds its way to the West. In implementing this policy, the United States attaches the highest importance to Turkey as the principal transportation corridor from Central Asia to Western Europe (Arvanitopoulos 1998). US policy towards Central Asia achieved its most significant success with the completion of the Baku–Tbilisi–Ceyhan oil pipeline crossing Azerbaijan, Georgia and Turkey, transporting Azeri oil over 1,760 km to the Turkish port of Ceyhan in the east Mediterranean.[11]

Russia

Russia continues as an active player in the geopolitics of Central Asia. Russia's principal aim in Central Asia is to maintain its influence in the region on the basis of, in Putin's words, 'good neighbourly relations and strategic partnership'. Specifically, besides maintaining control over the region's assets such as hydrocarbons and minerals, Russia would be anxious to counter the influence of Islamist extremist forces. The success of the BTC oil pipeline which will carry Azeri (and later Kazakh) oil Westwards to Europe, outside the Russian-controlled pipeline network, can be seen as fuelling Russia's concerns in the region. However, Russia's preponderant presence and its long-term role in the politics and economics of the region cannot be ignored; an American scholar noted this in April 2006:[12]

> Central Asian energy reserves are vast, but we should not exaggerate the role they are likely to play in meeting U.S. and Western needs … No matter how enlightened, U.S. policy will only have a marginal effect on minimizing Russian or Chinese presence in the region, as geography (even without the addition of geopolitical pressure) gives each more leverage.

Iran

Iran's geographical position provides an excellent opportunity for oil and gas pipelines to run from the Caspian Sea to the Persian Gulf and the Gulf of Oman. Iran's 90 years of experience in the oil and gas industries and its existing oil and gas installations, port facilities, refineries and networks of oil and gas pipelines also offer considerable advantages to Caspian oil and gas exporters. Iran's gas pipeline networks are already connected to the Republic of Azerbaijan and they are within a short distance of Turkmenistan. A pipeline connecting Kazakhstan and Turkmenistan to this network would be at least four times shorter and much cheaper than any of the proposed lines to other places such as the Black Sea and the Mediterranean.

Throughout the 1990s, Iran made major efforts to engage with Central Asian countries, particularly Azerbaijan and Turkmenistan, to pursue hydrocarbon development and transportation proposals. However, US opposition ensured that these efforts were unsuccessful and Iran was totally excluded from the principal projects. Under strong pressure from the US, Iranian companies were deliberately excluded from participating in the BTC oil pipeline project. The situation changed after 9/11. The subsequent war on terror launched by President Bush, and the inclusion of Iran into the 'axis of evil' changed Iran's approach to Azerbaijan. Fearing that Azerbaijan could host American military bases in a possible conflict with Iran, the Iranian government decided to engage in a dialogue with the Azerbaijani leadership; high-levels visits were exchanged to consolidate bilateral ties. Other issues, such as the opening of the Azerbaijani consulate in Tabriz, and the beginning of airline flights between Tabriz and Baku, were also quickly resolved (Cornell and Ismailzade 2005).

Other players

The geopolitics of the region are affected by other regional players, such as China, India, Iran and Turkey, as also by developments within Afghanistan and the role of outside players in this country. The Central Asian republics are now sovereign and independent countries with political aspirations of their own and the policies of these countries are guided by several considerations. They include the desire not to be dominated by any single regional or global power; to avoid entering into an exclusivist alliance with any power; to maintain as their primary focus raising the living standards of their people; and to promote political and economic cooperation amongst themselves to achieve their interests.

Conclusion

In West Asia, with about two-thirds of the world's oil reserves, there are new levels of violence, hate and destruction in Iraq, Turkish–Kurdish skirmishes in North Iraq, a near-civil war in Gaza, strong military action by Lebanese forces against Islamist elements, and an on-going confrontation between the US and Iran, centred around the latter's nuclear ambitions, which threatens to escalate to war. All of these conflicts are taking place amidst a resurgence of extremist Al-Qaeda-oriented forces. While Central Asia does not have a large-scale ongoing armed conflict, there is a chill in US–Russia relations following US plans to set up missile bases in East Europe (and recent Russian military action in Georgia), while the energy-related rivalry between them continues (Bhadrakumar 2007). Official US sources have in recent months been expressing concerns about China's expanding military strength, which is seen as upsetting the military balance of power in the region and encouraging an arms race.

Not surprisingly, this environment of insecurity, distrust and big power contention has led to what the Indian strategic affairs commentator, Brahma Chellaney, has described as a 'qualitative re-ordering of power' globally but most conspicuously in Asia, in which the major players are seeking to obtain the maximum possible geopolitical advantage for themselves on the basis of 'new equations and initiatives' (Chellaney 2007). Thus, the trilateral Russia–China–India dialogue and the evolving Shanghai Cooperation Organization are attempts by major Asian role players to explore and pursue common strategic, primarily energy-related, interests.

Separately, the USA, Japan, Australia and India took the first step in giving shape to a new 'Quadrilateral Initiative', when officials of the four countries met on the sidelines of the ASEAN Regional Forum (ARF) Summit in Manila in late May 2007. Concerned about China's sensitivities, the four countries have made it clear that they are not seeking to create a new security alliance but are only 'looking at issues of common interest' (Varadarajan 2007). While Indian spokesmen have defended the initiative by pointing out that there are 'many overlapping structures in Asia', Chinese commentators have expressed concerns about it, particularly about the Indian participation in the grouping. Thus, the People's Daily saw in the burgeoning naval cooperation between the four countries 'a signal for a new balance of force in the Asia region'. The United States, it said, was 'an old-brand power' but it is 'striving to win the support of Japan and India in a bid to prevent China and Russia from joining forces' (Varadarajan 2007).

The present Asian scenario suggests that, while a number of different interests and alliances are being pursued, no firm and enduring security architecture has been put in place that would bring together like-minded countries in opposition to another country or grouping. This is confirmed by the fact that every member of the 'Quadrilateral Initiative' has very substantial political and economic ties with China. Again, while the US, in tandem with Japan and Australia, is strongly pursuing relations with India, it is also maintaining substantial links with China and Pakistan.

It is interesting to note that, amidst the prevailing cacophony and uncertainty, China and India, which had been seen by many observers as competitors in the pursuit of their energy security interests, are building up bilateral ties on the basis of cooperation in the energy sector. Former Indian Petroleum Minister Mani Shankar Aiyar, in public remarks in Beijing, in January 2006, laid the foundation of an energy-based relationship between them in the following terms (Aiyar 2006):

> It is in search of such mutual cooperation to the mutual benefit of our two countries that I have come to Beijing … *to engage China and India in a strategic and cooperative quest for energy security. For our part, we look upon China not as a strategic competitor but as a strategic partner. Our cooperation*

> *in energy is based on equal cooperation, mutual benefit, mutual respect and enhanced understanding.* [Emphasis added.]

He concluded his speech with these words:

> China has an excellent record of regional cooperation in Asia. So does India. Together, we can set the agenda for Asian energy cooperation. That is how the *Asian Quest for Energy Security could lead to Asia regaining its traditional place – a place it has held for thousands of years of recorded history and lost only in the last two hundred years or so – in the vanguard of the advancement of human civilization.* The Asian Renaissance brought us all to independence and liberation. Now, the Asian Resurgence depends on energy cooperation in Asia. The 21st century will indeed be the Asian century only if Asian countries – buyers or sellers – join hands together in a continent-wide bid at bringing Asia together and keeping Asia together. I am confident that we will. [Emphasis added.]
>
> (Aiyar 2006)

India has just commenced a broad-based, constructive energy-related engagement with West Asia and Central Asia. It is emerging as a significant player in the international and regional energy economy and is increasingly being solicited as a partner in the global quest for energy security. The challenge before the country is to understand the importance and implications of the global and regional developments set out above and effect the required adjustments in its politics and policies and, above all, in its mindset, in order to respond effectively to them.

Notes

1 This chapter is a revised version of the paper presented at the TERI-KAF Conference on 'India's Energy Security: foreign, trade and security policy contexts', 29–30 September 2006.

2 'Persian Gulf Oil and Gas Exports Fact Sheet', Energy Information Agency [EIA], Department of Energy, Washington DC, April 2003.

3 Energy security has attributes that distinguish it from other aspects of national security: first, while various aspects of national security are generally 'status-quo-ist' in that they protect and sustain the existing order, be it national borders, national political structures or national values, energy security is a dynamic concept in that it enhances a nation's economic and, therefore, political status by providing it with the resources to pull its people out of poverty and pursue national growth and development.

4 In response to this challenge, the international oil industry is already integrating in significant ways: major companies are merging to pool together their financial resources and technological capabilities. Again, there is a clear trend in favour of national oil companies integrating across the hydrocarbon value-chain, from exploration to production, to transportation, to refining and petrochemicals. Finally, E&P contracts in developing producer countries are increasingly being linked to refinery proposals and, on occasion, even to other infrastructure development proposals such as roads, railways, power, mining and port development projects.

5 Delhi Declaration, signed by King Abdullah bin Abdul Aziz Al-Saud of Saudi Arabia and Prime Minister Dr. Manmohan Singh of India, on 27 January 2006.
6 Though in money terms the Indian presence is relatively modest, India has established itself in sectors such as pharmaceuticals, tea, readymade garments, woollen goods, leather goods, as jute manufacturers, cosmetics, cotton yarn, machinery, machine tools, rice, plastic products, machinery and instruments, electronic goods and chemicals.
7 'China, Russia, India hold trilateral summit' in The People's Daily Online, 18 July 2006. Online at: english.people.com.cn/200607/18/print20060718_284229.html
8 Azerbaijan remains locked in disputes with Turkmenistan and Iran over competing claims to overlapping petroleum fields. Tehran maintains that the 1921 and 1940 agreements between the Soviet Union and Iran on the Caspian Sea are valid until a new legal regime has been agreed to by the five littoral states. Iran, meanwhile, regards any unilateral or bilateral deals for hydrocarbon exploration of the Caspian as being null and void. Most of Azerbaijan's oil resources (proven as well as possible reserves) are located offshore, and perhaps 30–40 per cent of the total oil resources of Kazakhstan and Turkmenistan are offshore as well. Overall, proven gas reserves in the Caspian are estimated at 177–182 TCF. Possible reserves could yield another 293 TCF of natural gas.
9 'Iraq Oil Law gets Cabinet Approval as Federal Vision Wins Out for Now', *Middle East Economic Survey*, 49(10), 5 March 2007.
10 'Kazakhstan Plans Downstream Projects with Russia, Turkey', *Middle East Economic Survey*, vol. 49(22), 29 May 2006.
11 The transportation of Central Asian oil to Ceyhan is not just meant for Western markets: there are plans for this oil to be transported to Asian markets as well through an undersea pipeline covering the 400 km distance from Azerbaijan to Ashkelon in Israel, and to pump this oil to Eilat on the Red Sea through the Ashkelon–Eilat pipeline; from Eilat, oil tankers would be in position to take this Caspian oil to India and further afield to South-East Asia and North-East Asia.
12 'Kazakhstan Plans Downstream Projects with Russia, Turkey', *Middle East Economic Survey*, 49(22), 29 May 2006.

References

Abi-Aad, N., 2002. 'Will Central Asia be an alternative to the Gulf?' *Middle East Economic Survey*, 45(18), p. D5.

Aiyar, M. S., 2005. Inaugural address at the 'Round Table of Asian Ministers on Regional Cooperation in the Oil Economy Stability, Security, Sustainability Through Mutual Interdependence', 6 January, 2005, New Delhi. In *A historic beginning for Asian oil economy* (proceedings of the First Round Table), Ministry of Petroleum and Natural Gas, Government of India, pp. 10–11.

Aiyar, M. S., 2006. 'Asia's quest for energy security.' Text of speech by Minister of Petroleum and Natural Gas, Beijing, 13 January, 2006, in *Frontline*, 23(3), Feb 11–24, 2006, viewed 30 May, 2008 <http://www.flonnet.com/fl2303/stories/20060224002309000.htm>.

APS, 2006. 'The Caspian Sea – the history and legal background.' *APS Review*, 34th Year, 67(1), July 3–10, 2006.

Arvanitopoulos, C., 1998. *The geopolitics of oil in Central Asia*, Thesis, Winter 1998, viewed 1 June, 2008 <http://www.hri.org/MFA/thesis/winter98/geopolitics.html>.

Bhadrakumar, M. K., 2007. 'US gives Russia short shrift.' *Asia Times*, 19 June 2007 viewed 1 June, 2008 <http://www.atimes.com/atimes/Central_Asia/IF19Ag01.html>.

Chanlett-Avery, E., 2005. 'Rising energy competition and energy security in Northeast Asia: issues for US policy.' *CRS Report for Congress*, updated 9 February, 2005, CRS 18.

Chellaney, B., 2007. 'New great game.' *The Asian Age*, New Delhi, 2 June, p.7.

Cornell, S. E. and Ismailzade, F., 2005. 'The Baku–Tbilisi–Ceyhan pipeline: implications for Azerbaijan.' In Starr, F. and Cornell, S. (eds), *The Baku–Tbilisi–Ceyhan pipeline:*

window to the West. The Central Asia-Caucasus Institute and Silk Road Studies Program, Washington DC.

Dixit, J. N., 2004. 'Emerging international security environment: Indian perceptions with focus on South Asian and Central Asian predicaments.' In Santhanam K. and Dwivedi, R. (eds), *India and Central Asia – advancing the common interest.* New Delhi: Anamaya Publishers, pp. 19–20.

ECSSR, 2007. 'Economic security issues and the role of the Gulf energy sector.' In *Gulf oil and gas - ensuring economic security*. Abu Dhabi: The Emirates Center for Strategic Studies and Research, p.11.

EIA, 2007. *Country analysis briefs: Caspian Sea*. Energy Information Administration, US Department of Energy, Washington D C, viewed 30 May, 2008 <http://www.eia.doe.gov/emeu/cabs/Caspian/pdf.pdf>

Ernst & Young, 2006. Presentation at the Ministerial Round Table on Cooperation between North and Central Asian Producers and Principal Asian Consumers regarding Stability, Security and Sustainability in the Asian Hydrocarbons Economy, 25 November 2005, New Delhi. In *New vistas for regional cooperation in Asian Oil economy* (proceedings of the Second Round Table), Ministry of Petroleum and Natural Gas, Government of India, p. 94.

Exim Bank, 2006. 'GCC countries: a study of India's trade and investment potential., *Research Brief*, No. 23, March 2006.

Fesharaki, F., 2005,.Presentation at the Round Table of Asian Ministers on Regional Cooperation in the Oil Economy Stability, Security, Sustainability Through Mutual Interdependence, 6 January 2005, New Delhi. In *A historic beginning for Asian oil economy* (proceedings of the First Round Table), Ministry of Petroleum and Natural Gas, Government of India, p.101.

GOI, 2001. 'Report of the group on India hydrocarbon vision–2025.' In Jasjit Singh (ed), *Oil and gas in India's security*. New Delhi: Knowledge World, in association with IDSA, pp. 131–230.

GOI, 2005. *A historic beginning for Asian oil economy* (proceedings of the Round Table Asian Ministers on Regional Cooperation in the Oil Economy Stability, Security, Sustainability Through Mutual Interdependence), Ministry of Petroleum and Natural Gas, Government of India.

Janabi, A., al-, 2004. 'Oil policy options for Iraq.' *Middle East Economic Survey*, 47(9), p. D3.

Leverett, F. and Bader, J., 2005. 'Managing China-US energy competition in the Middle East.' *The Washington Quarterly* 29(1), p. 188.

Massoud, A., 2006. 'Oil may fuel Sino-US conflict.' *Al-Jazeera*, 29 June.

Mekay, E., 2007. 'Selling Iraq by the barrel.' *Asia Times*, 2 March, 2007, viewed 30 May, 2008 <http://www.atimes.com/atimes/Middle_East/IC02Ak04.html>.

McKinsey & Company, 2005. Presentation at the Round Table of Asian Ministers on Regional Cooperation in the Oil Economy Stability, Security, Sustainability Through Mutual Interdependence, 6 January 2005.,Iin *A historic beginning for Asian oil economy* (proceedings of the First Round Table), Ministry of Petroleum and Natural Gas, Government of India.

Olcott, M. B., 2005. 'Russia, Central Asia and Caucasian threats: a four year assessment.' Testimony prepared for the House Committee on Armed Services (HASC) Threat Panel Hearing on Threat In Eurasia, 22 September, 2005, Carnegie Endowment for International Peace, viewed 1 June, 2008 <http://www.carnegieendowment.org/files/OlcottTestimonySept2005.pdf>.

PEL, 2006. Presentation at the Ministerial Round Table on Cooperation between North and Central Asian Producers and Principal Asian Consumers regarding Stability, Security and Sustainability in the Asian Hydrocarbons Economy, 25 November 2005, New Delhi. In *New vistas for regional cooperation in Asian oil economy* (proceedings of the Second Round Table), Ministry of Petroleum and Natural Gas, Government of India, p. 84.

Planning Commission, 2006. *Integrated energy policy: report of the Expert Committee.* Government of India, p. xiii.

Roy, M. S., 2006. 'The Shanghai cooperation organization: a critical evaluation.' IDSA strategic comments, Institute of Defence Studies and Analysis, New Delhi, 4 July 2006, viewed 29 May, 2008 <http://www.idsa.in/publications/stratcomments/MeenaSinghRoy040706.htm.

Varadarajan, S., 2006. 'Energy key in the new Asian architecture.' *The Hindu*, New Delhi, 25 January.

Varadarajan, S., 2007. 'Four-power meeting drew Chinese demarche.' *The Hindu*, New Delhi, 14 June p.15.

Wu Lei, 2006. 'China-Arab energy cooperation: the strategic importance of institutionalization.' *Middle East Economic Survey*, 49(3), p. 28.

6 Natural gas pipelines

Geopolitics, affordability, security dimensions

Ravi Kumar Batra

India is currently the fourth largest energy consumer in the world. With its GDP projected to grow at eight per cent per annum, energy demand is expected to rise to around five to six per cent per annum, which, in effect, means doubling of current consumption in about 12 to 13 years. While coal has been the major component of India's energy basket, gas, which is currently at around nine per cent, could well grow to about 15 to 20 per cent, if India is able to meet demand. Gas supply is currently around 100 MMSCMD[1] which includes both domestic gas and imported liquefied natural gas (LNG) and once gas is freely available the demand for the fuel is likely to rise appreciably. Presently, the fertilizer and power sectors are the major consumers of gas at 29 per cent and 40 per cent respectively with the balance being consumed in the industrial and domestic sectors. With large gas discoveries in the KG Basin off India's south-eastern coast, the domestic gas position is likely to improve considerably but the need for imports will continue.

In addition, India is now importing increasing quantities of coal and acquiring coalmines abroad. Though, under the Kyoto Protocol, the country is not currently required to cap its greenhouse gas emissions, there is already considerable pressure on India to reduce its carbon footprint. Natural gas is therefore a welcome substitute for coal in the power and industrial sectors.

India already has three LNG terminals on its west coast and is constructing a fourth one which will be commissioned by 2012. Three transnational pipeline proposals, listed below, have been mooted from time to time but for various reasons have either stalled or are making slow progress. These include the Iran–Pakistan–India gas pipeline, the Myanmar–India gas pipeline and the Turkmenistan–Afghanistan–Pakistan–India gas pipeline. Furthermore, the Oman–India sub-sea gas pipeline, an idea once considered as dead as a dodo, has been resurrected.

The largest economic power in the region, China, has been looking to source gas from some of the same gas-rich countries as India and has been far more successful, albeit for different reasons. The various pipeline proposals for gas supply to India and pipeline projects of China that have impacted, or could impact India, are illustrated in the map (Figure 6.1).

We describe, in this section of the book, the status of each pipeline project along with pending issues that need to be addressed such as gas reserves, pricing and

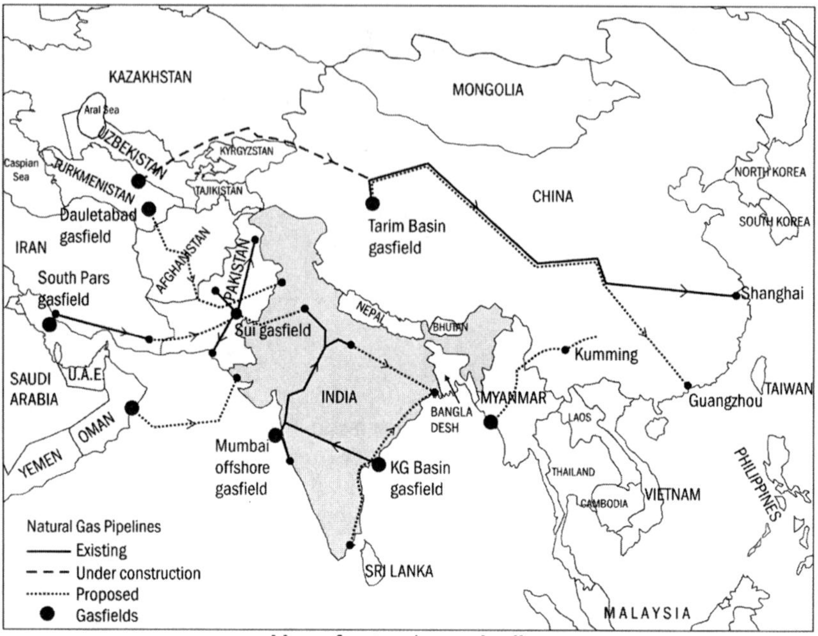

Map of natural gas pipelines

Figure 6.1 Natural gas pipelines.

energy security. The way China has gone about securing gas through transnational pipelines offers a sharp contrast to that of India and is discussed in some detail.

The Iran–Pakistan–India gas pipeline

The Iran–Pakistan–India (IPI) gas pipeline was first mooted by Dr R K Pachauri, Director General, TERI, India and Dr A S Ardekani of Iran as far back as 1989. Iran has the second largest gas reserves in the world and it therefore made eminent sense to supply gas to India. However, the main problem as seen at that time, was that Pakistan would be a transit country and relations between India and Pakistan were never uniformly cordial, to say the least. The main worry was that Pakistan could interfere and stop the flow of gas to India, thus holding India hostage. This was notwithstanding the fact that the Indus Water Treaty between two countries had operated very smoothly; in fact to the extent that in the public domain there was hardly any discussion on it and it was largely forgotten. However, it was the rocky relationship between the two countries, primarily on the issue of Kashmir, which stymied any progress for the next 15 years.

Matters took a positive turn after Pakistan discovered that it would soon be running short of gas thanks to the extensive development of its natural gas market and its own limited domestic resources. Thereafter, there was a revival of interest including the first detailed assessment of the project by the Anglo-Australian company BHP Billiton in the year 2003. As conceived by BHP Billiton, a 44-inch pipeline was to be laid from Asalouyeh on the Iranian coast where gas from the South Pars field would be pumped 1,115 kilometres across Iranian territory to the Pakistan border and a further 760 kilometres through Pakistan to the Indian border. At a distance 70 kilometres short of the Indian border the pipeline would link up with Pakistan's own gas network, the Sui Northern Gas Pipeline Ltd. (SNGPL) as well as its southern counterpart before entering India. Within India a further 600 kilometres would be required to connect with the major west-north transmission line. It was anticipated that at its full capacity Pakistan would use about 60 MMSCMD and India 90 MMSCMD. The total cost of the project was estimated at $4 billion.

Arising from these developments, India's security concerns abated as Pakistan now had a stake in the smooth operation of the pipeline and India's formal commitment to the project was made by the Prime Minister, Dr Manmohan Singh, in September 2004. Despite the unrest in Baluchistan where Pakistan's own gas lines from the Sui gasfield were regularly blown up by insurgents, it was felt that the project could go forward, provided certain measures were taken to physically protect the pipeline and, if breached, to repair it quickly and at short notice. India has had its own crude oil pipelines blown up in Assam from time to time and though inconvenient, the situation has been manageable.

Since 2004, while discussions on a bilateral or trilateral basis between the three countries made some progress, there have also been road blocks and changes in the base numbers and costs. First BHP Billiton revised the cost to around $7 billion because of increases in steel prices, diameter of the pipeline etc. Then it was rumored that Pakistan was demanding a transit fee of as high as $700 million, though this was subsequently denied. Meanwhile, a contract that India had entered into for supplies of LNG from Iran came unstuck because Iran claimed that it had not been given the final seal of approval by their Supreme Economic Council. The fact that India at the same time cast its vote against Iran in the IAEA on the nuclear issue and gas prices had gone up substantially in the international market was not mere coincidence. This brings into question the sanctity of contracts concluded with Iran, especially as the pipeline contract, despite any commercial agreement that may be reached between corporate entities, will again need to be approved by the Supreme Economic Council. Iran is not the easiest country with which to do business.

The indicated price of gas at the Indian border is around $5/MMBtu[2] against an earlier expectation of around $2/MMBtu, excluding transit and transportation fees payable to Pakistan. However, this price was discussed in 2004 at a time when crude oil prices were around $60 a barrel. Whether Iran will ask for a higher price (in real terms) now that crude oil is upward of $100 a barrel remains to be seen. Iran has proposed that gas prices be revised every three years; something that India, in particular, is unhappy about.

Iran has proposed that each country build its own section of pipeline; not a very satisfactory arrangement in terms of operation of the entire network, obtaining international finance and gaining the confidence of customers in terms of reliability of supply. Despite a year having passed, in a meeting in May 2008

India and Pakistan could not come to an agreement on transit and transportation fees. Meanwhile, there has been a change on the Iranian side. It has laid a pipeline, IGAT-7, for domestic use from Asalouyeh to the eastern town of Iranshahr, which will be extended 100 kilometres to the Pakistan border to meet Pakistan and India's requirements. The spare capacity of this pipeline is 60 MMSCMD and therefore the availability stands reduced to 30 MMSCMD for each country. In the case of India this is one-third of what was originally planned and considerably diminishes the importance of the pipeline, unless its capacity can be augmented later to restore the supply to the original figure. Strangely, through all these developments, the cost of the pipeline continues to be touted at $7 billion despite its length in Iran having been reduced by about 1000 kilometres and for a much lower throughput.

In constructing the pipeline, Iran's contribution will be restricted to extending the IGAT-7 pipeline by just 100 kilometres. Pakistan will carry the biggest risk as the capacity of its section of the pipeline will need to meet India's as well as its own requirement i.e. 60 MMSCMD.

On a visit to China in early 2008, President Musharraf of Pakistan suggested that China could be a partner in the project, if India backed out. Finally, there is the American view, not favouring the IPI pipeline as it is felt that the income will help assist Iran's alleged nuclear weapons programme. Much will depend on how firm India stands on this issue. For whatever reason, in the event the pipeline does not come through, Iran and Pakistan will need to re-evaluate the economics of Pakistan being the sole buyer, unless China steps in, and takes India's place.

Myanmar–India gas pipeline

The A1 and A3 blocks in the Rankhine offshore area of Myanmar, near the port of Sittve, have gas reserves between 160 to 280 BCM, as estimated by Gaffney Cline and Associates. India's share in these blocks is 30 per cent, with a 20 per cent participating interest held by ONGC Videsh Ltd. and 10 per cent by Gas Authority of India. Kogas of Korea holds 10 per cent while Daewoo of South Korea is the operator with 60 per cent interest.

In February 2004, a letter of intent was issued to GAIL making it the preferential buyer of gas from block A1. Subsequently, in March 2006, an MoU was signed, reinforcing that GAIL would be a preferential buyer of natural gas from Myanmar but at the same time a similar MoU was signed with China. Earlier, in January 2005, India had signed a bilateral agreement with Myanmar to import gas through Bangladesh. However, Bangladesh insisted that other bilateral issues be discussed with India, including trade imbalances and transit of electricity to which India did not agree. As a result, Bangladesh did not allow the pipeline to be laid through its territory. India then examined the possibility of bypassing Bangladesh by laying a much longer line from Myanmar through India's north-eastern states. GAIL separately also examined the possibility of compressing the gas (CNG) and moving it in ships to an Indian port. Around that time Myanmar also invited bids for export of the gas in the form of LNG but the proposal was not acted upon with any great seriousness as the gas reserves did not justify this option.

In January 2006 Myanmar advised India of China's intention to buy the gas and in February 2007 it was agreed that the entire production of 16 MMSCMD of gas from the A1 and A3 fields would be sold to PetroChina at a wellhead price of $4.279/MMBtu. China is to lay a 2,380 kilometre pipeline connecting the gas field to Kunming in China and beyond. This is reminiscent of the temporary oil pipeline laid during World War II from the port of Budge Budge, south of Calcutta, to Kunming. China is also to pay $150 million annually for 30 years as a transit fee for the 990 kilometre stretch through Myanmar. There was considerable dismay in India at this development and an inter-ministerial spat took place on how the whole issue had been handled over the years.

The award of the gas contract to China was attributed to China having vetoed a draft resolution tabled in the UN Security Council to ease repression and release political prisoners in Myanmar. Though this may well be the case, there is no doubt that India lost considerable time looking at various options, carrying out feasibility reports etc. Equally, it is not difficult to see that this contract has strategic and security value for China as, apart from the gas from Myanmar, the pipeline could, at some future date, be used to ferry re-gasified LNG by building an LNG terminal at Sittve. It also opens up the possibility of laying a parallel oil pipeline to transship oil from the Middle East and avoid the long haul through the Straits of Malacca. It is not surprising that China pulled out all stops to secure this contract.

Turkmenistan–Afghanistan–Pakistan–India gas pipeline

The first to propose a gas pipeline from southern Turkmenistan through Afghanistan to Pakistan was the Argentinean company, Bridas, in 1992. However, due to US pressure the project was re-allocated in 1996 to an American company, Unocal, which later pulled out of the project because of the very unstable conditions in Afghanistan. The project was revived in 2002 when the heads of state of the three countries signed an agreement to implement the Turkmenistan–Afghanistan–Pakistan (TAP) project. Gas from the Dauletabad field was to be fed through a 56-inch diameter pipeline, 1,680 kilometres in length, and with a capacity of 90 MMSCMD, to Pakistan via Herat and Kandahār in Afghanistan to join up with Pakistan's domestic gas network. The Asian Development Bank (ADB) was appointed the lead development partner to facilitate the project, to serve as an honest broker and as the lender of last resort. The cost of the project was pegged at $3.3 billion. A number of steering committee meetings were held under the aegis of the ADB, but details of the project could not be finalized. It was felt that for the project to be viable, it should be extended to India to link up with the transmission line feeding gas to northern India. India agreed to be an observer and only recently decided to become a member, after which the project was renamed as the Turkmenistan–Afghanistan–Pakistan–India (TAPI) pipeline project. The cost has now escalated to over $7 billion. India's stated reluctance has been that it was not sure the gas reserves in Dauletabad could service the requirements of all three countries over a period of 30 years. The unstated reasons though have as much to do with the politically unstable situation in Afghanistan and related security fears.

The reserves feeding the TAPI pipeline were originally estimated at 1.7 TCM by Russian geologists. In an audit carried out by DeGolyer & MacNaughton along with Gaffney Cline in 2003, the estimated reserves were raised to 4.5 TCM, more than sufficient to meet a demand of 90 MMSCMD over 30 years. Nonetheless, in 2008 Gaffney Cline was asked to make an independent assessment of all of the country's gas reserves. Six years have thus passed with very little progress.

The contrast with the manner in which China has gone about securing gas supplies from Turkmenistan has been stark. In 2003 China first negotiated with Kazakhstan a gas pipeline that would follow the alignment of the existing oil pipeline. This became known as the Central Asia China gas pipeline. In 2006, Turkmenistan agreed to join in supplying gas to China through this pipeline and Uzbekistan followed shortly thereafter in 2007. The price of gas is expected to be around $4.00 per MMBtu at the Turkmenistan/Uzbekistan border. Starting from the Bagtyarlyk gasfield on the right bank of the Amudarya River in Turkmenistan near the border with Uzbekistan, two branches will be routed through Kazakhstan and Uzbekistan. They are to join up when they enter west China and in 2009 link up with the existing West East gas pipeline that starts from the Tarim Basin and feeds Shanghai and other markets to the extreme east. The Chinese have independently verified the volume of reserves at Bagtyarlyk at 1.3 TCM. Apart from building the pipeline, the Chinese company CNPC is providing financing and technical know-how for the gas processing and purification facilities, compressor stations etc. In a second phase the capacity will be increased to 80 MMSCMD to be completed by 2010. As the capacity of the West East pipeline is 30 MMSCMD, a parallel pipeline is likely to be laid which will not only feed the area around Shanghai but also Guangzhou in south eastern China. The exact alignment of the pipeline is, however, not known. The entire pipeline will be about 9,000 kilometres long and will supply China 80 MMSCMD of gas for 30 years. In a relatively short time frame, China has not only struck deals to buy gas from the three central Asian countries but will also have built this long-distance pipeline. The speed with which it has moved has been quite impressive.

Currently, the right bank of the Amudarya River is reserved for China. Iran will continue to receive volume from the general area of Korpeje. Afghanistan, Pakistan, and India will get gas from Dauletabad. The Russian company Gazprom exports 140 MMSCMD of Turkman gas annually to Europe. The commitment to supply China would raise total exports to 220 MMSCMD. This excludes Turkmenistan's own domestic gas consumption of 50 MMSCMD. Doubts have arisen as to the ability of Turkmenistan to feed its export markets in the short term. Besides, if the TAPI pipeline does not make quick progress, it may find that Dauletabad gas gets increasingly diverted to Russia to feed the growing west European markets.

Oman–India sub-sea natural gas pipeline

In 1995, the Oman Oil Company headed by the maverick John Duess proposed laying a 1,100 kilometre sub-sea pipeline of 24-inch diameter that would link Oman gasfields to a landfall point in the state of Gujarat in India. Much excitement was generated by this proposal on account of the short distance and the fact that no transit through a third country was involved. However, the proposal could not materialize as, first, there were technology problems manufacturing a pipe thick enough to withstand the water pressure at depths of 3,500 metres; second, no submersible vessel had been developed to repair the pipeline at that depth if required; and third, the strong ocean currents in the seabed (the Indus fan) could affect the stability of the pipeline. Finally, India felt that the reserves of gas in Oman would not be sufficient to sustain delivery over a period of 30 years.

After 13 years the project has now been revived by the South Asian Gas Enterprises (SAGE) whose key team members were associated with the original project. SAGE has confirmed that the earlier technical difficulties have now been overcome. Corus, the UK Steel Company recently acquired by Tata Steel of India, now has the technology to manufacture steel pipeline for the project. It has won a contract to supply a 312-kilometre length of pipeline of the required specification, to be laid on the seabed in the Gulf of Mexico for the Perdido North Project at depths of up to 3,900 metres. Heerema, the Netherlands-based marine engineering contractor, has developed a new barge that can lay pipelines at that depth. SAGE maintains that no deepwater large diameter pipeline has ever required in-situ repairs. Nonetheless remote operated repair systems have been developed over the last five years, but will take time to carry out repairs. SAGE considers that by providing two lines, with a built in redundancy factor, the need for repairs will not arise. SAGE has therefore proposed to lay two 28-inch lines, each of which would carry 31.5 MMSCMD gas to India. The project cost is estimated at between $2.1 to 3.4 billion for the first line, which could deliver gas by 2012. SAGE has suggested that the gas sellers and/or buyers could be the owners of the pipeline and could set the tariff, which is expected not to exceed $1.30/MMBtu for the first line, dropping to $1.10/MMBtu with the commissioning of the second line.

The technical feasibility of the proposal as well as the cost estimates will need to be addressed in detail. Presentations have been made to various ministries of the Government of India. If feasible, this may address several of the security and transit issues that India faces with regard to overland pipelines. However, because of the earlier concern regarding Oman's gas reserves there will need to be land-based gas pipeline linkages on the Arabian Peninsula with other countries that can augment the supply of gas to India through this route.

Analysis and conclusions

Though India is surrounded by countries that have abundant gas reserves, it has not been able to tie up a single project in over a decade. Geopolitics has played a dampening role and the security of the pipelines and of gas supply has been an important factor. Infrequent meetings with long gaps with the concerned countries haven't helped.

India has dragged its feet for over a year on the Iran–Pakistan–India gas pipeline and even now is not happy with the proposed transportation and transit fees proposed by Pakistan. While the nuclear deal with the USA was being negotiated, India seemingly did not want to offend the USA and therefore stalled discussions on the pipeline by highlighting but not moving forward on these two issues which, in terms of the total gas supply cost, were relatively minor items. China's lately expressed willingness to come onboard and become a partner in the proposal has been welcomed mainly by Pakistan but also by Iran. While some see this as an effort to get India to move faster in the matter, the China factor needs to be taken more seriously. Considering the extent to which it is prepared to pipe gas over long distances, as in the case of Turkmenistan and Myanmar, it is not inconceivable that a pipeline could be laid from Pakistan along the Karakoram highway and through the Karakoram Pass into western China. China would only be too pleased to get access to the vast gas reserves of Iran.

In the case of the Myanmar pipeline, India got bogged down looking at various options and was perceived as a slow mover. Also, it would appear that complacency set in with GAIL being initially nominated as the sole marketer of the gas. China did not take long to step in and, for the reasons discussed earlier, was able to wrest the project away from India.

India has reluctantly and recently come on board the Turkmenistan–Afghanistan–Pakistan–India pipeline project as it is commonly held that the writ of the Government of Afghanistan runs mainly in the capital city of Kabul and not in the badlands elsewhere. Of the two pipeline proposals from Iran and Turkmenistan, the former is more likely to fructify. Iran can easily supply the relatively small quantity of 60 MMSCMD to Pakistan and India; second, there is only one transit country for India as opposed to two for the Turkmenistan pipeline (the greater the number of transit countries the more complex the negotiations); third, Pakistan has an equal stake in the quantity of gas to be supplied by Iran; fourth, the newly elected democratic government in Pakistan is better placed than its predecessor to tackle and quell the unrest in the Baluchistan province through which the pipeline will pass. Finally, once the pipeline is operating smoothly it should also be possible to increase capacity. For this to happen India will need to move more decisively and quickly than it has in the past to sew up the deal, unmindful of US pressures. In trying to keep all parties happy it may end up displeasing everyone. If India has a pipeline strategy at all, it is not readily apparent.

The Oman–India pipeline proposal is a resurrection of an earlier one. As it is one of the world's deepest sub-sea proposals, India will need to be satisfied as to the proposed technology as well as the competence of South Asian Gas Enterprises and associated organizations. If these are clearly established, the project has the best chance of being implemented as the economies look good, no transit is involved and security risks are minimal.

One of the unnoticed advantages of receiving gas from Iran, Turkmenistan and Oman is that it will result in a more balanced internal distribution of gas in India than would otherwise be the case, leading to savings in transportation costs. Currently it is planned to move domestic KG Basin gas all the way to the north. Gas from Iran and Turkmenistan could feed the north, LNG terminals and Omani gas could feed the west and south with the KG basin gas feeding the west, south and the east.

India needs to accept that there is an element of risk inherent in the petroleum sector as exemplified in the area of exploration for crude oil and gas and even in the marketing of petroleum products, where the imposed cap on pricing of transportation fuels has left the state owned oil companies bleeding. Pipelines carry a smaller risk and can operationally and financially be structured to ensure that any disruption in supply does not carry a heavy penalty. Sabotage of a land-based pipeline can, to a large extent, be prevented; first, by burying the pipeline well below the ground; second, by having an optical fiber sensing facility to detect any break or leak in the line; third, by an overhead satellite monitoring system; fourth, by pipeline patrolling. Positioning repair teams at regular intervals and limited underground gas storage can take care of short-term supply disruptions.

One perplexing aspect of all the long pending pipeline proposals has been the diffidence of the Government of India to negotiate upfront the well-head price of gas, while spending an inordinate amount of time on gas reserves, pipeline alignments, transportation and transit fees, and security issues. With no commitment on prices, the supplying countries have enjoyed the benefit in the last few years of rising international gas prices. Also, increases in steel prices have led to doubling of the cost of pipeline infrastructure. What has held India back is; first, the low price of domestic gas, which makes market-based prices of gas look expensive; second, the inability of the power sector to raise electricity prices, instead being required to provide cheap power to the agricultural sector (thus leading to high commercial losses); and third, the higher subsidies that will be given to the fertilizer sector if gas prices are increased.

Gas at $5 or even $9 per MMBtu is considerably cheaper than oil at $100 a barrel ($17 per MMBtu). While India purchases, without demur, crude oil at prices ruling on the day, it seems to have a mental block in accepting that gas price trends will mimic crude oil prices while still remaining very attractive. India's long-term energy needs require that it accesses all energy forms from multiple sources to enhance its energy security. Success in importing gas by pipeline from neighbouring countries will depend on a recognition that the period of cheap energy is over and the opportunities that exist today may not be there tomorrow. India must seize the moment.

Notes

1 4 MMSCMD (million standard cubic meters a day) is required to feed a 1000 MW modern power station for one year.

2 Crude oil at $100 per barrel is equivalent to $17 per MMBtu (million British Thermal Units) in heat value).

References

Airy, A., 2005. 'Myanmar offers well-head gas at $2.52/MMBtu.' *Financial Express*, 18 August.

Airy, A., 2008. 'Pipe dream: India, Pak agree to disagree on transit, transport fees.' *Financial Express*, 7 May.

Analysis: 'China and Turkmen energy.' *United Press International*, 4 January, 2008.

Analysis: 'U.K. firm to audit Turkmen gas.' *United Press International*, 28 March, 2008.
Batra, R., 2007. 'Gas without borders.' *Hardnews*, 4(7), pp. 33–35.
Chatterji, S., 2008. 'Govt. goes slow on IPI pipeline.' Viewed 6 May, 2008 <http://www.business-standard.com>.
'CNPC starts building Tukmenistan–China gas pipeline.' 2008, *News Central Asia*, 25 March.
'India joins mega gas pipeline project.' 2008. *The Economic Times*, 25 April.
Mohan, N. C., 2008. 'Turkmenistan deal revives Iran hopes.' *Hindustan Times*, 26 April.
'Myanmar ditches India for China in gas deal.' *Times of India*, 9 April, 2007.
'Myanmar gas blocks award to China a setback for India.' *Economic Times*, 1 May, 2007.
'Myanmar gas for China: PMO wants petroleum ministry to try again.' *indianpetro.com*, 9 July, 2007.
Ranjan, A., 2007. 'Myanmar gas bid lost, MEA and Petroleum in war of words.' *The Indian Express*, 30 July.
Saleem, F., 'Pipelines or pipe dreams?' Viewed 27 June, 2004 <http://www.jang.com.pk>.
Siddiqui, H., 2005. 'Turkmenistan gas reserves inadequate.' *The Financial Express*, 18 February.
Subramanian, N., 2008. 'Accord on TAPI gas pipeline project.' *The Hindu*, 25 April.
'Text-Corus Tubes wins contract for pipeline.' *Reuters India*, 5 November, 2007.
Wan Zhihong, 2008. 'Building starts on gas pipe.' *China Daily*, 23 February.
Xiao Wan. nd. 'Gas up.' *China Daily*. Viewed 31 March, 2008 <http://chinadaily.com>.
'Yangon move shouldn't surprise oil ministry.' *Times of India*, 9 April, 2007.

7 India–China energy cooperation

Commonalities, synergies and complementarities[1]

Sudha Mahalingam

Decoupling energy consumption from economic growth is the foremost challenge facing the developing world, as much from the standpoint of environmental sustainability as from energy security. Despite concerted and fairly successful efforts to improve energy efficiency, both China and India continue to tread an energy-intensive growth trajectory relative to the developed world, thanks to their low growth base that looks to be the basis for quantum leaps in future consumption. China, by virtue of its large manufacturing base (necessitated and sustained by its emerging status as the pre-eminent supplier of manufactured goods to the United States[2]) is tied to an energy consumption pattern that closely trails its GDP growth. India, despite its skewed economy with a disproportionate share for the service sector, has also recorded energy-demand growth rates that shadow economic growth. India's demand is fuelled by a burgeoning and increasingly prosperous middle class attracted to the more energy intensive lifestyles sold by satellite television and overseas travel. While a striving for improved standards of living on the part of the middle class is understandable, it remains the case that India has a skewed consumption pattern, where incremental energy supplies are disproportionately consumed by the urban middle class while the remaining population continues to be energy starved.[3]

In a reflection of the growing importance of India and China in determining global energy consumption patterns, the 2007 World Energy Outlook report (published by the International Energy Agency) focused exclusively on these two countries (IEA 2007). Already, India and China together account for 45 per cent of incremental oil demand. IEA predicts global oil demand, which stands at 4,233 million tonnes per year today, will rise to about 5,780 million tonnes per year by 2030, primarily fuelled by China and India growing at a moderate pace – one which IEA calls the 'reference scenario' (IEA 2007). A higher growth rate will entail even higher oil-demand growth. Currently, both India and China have been clocking growth rates around eight to nine per cent, which translates into annual growth in commercial energy consumption of over five per cent compared to the world average of less than two per cent. In fact, India and China are at the vanguard of growth in global oil demand in this millennium, and consequently have a large role to play in the current spiraling oil prices.

Naturally, with such energy intensity, both economies are also carbon intensive. To China goes the dubious distinction of being the second largest emitter of greenhouse gases (GHG) after the United States, whereas India ranks fourth in the list. This is in spite of the fact that in per capita terms, both these economies are way below global average. Both countries are under pressure to switch to a more environmentally sustainable growth paradigm.

The similarities do not end there. Both countries are big oil importers. India imports almost three out of every four barrels of oil it consumes, while China's imports have grown from zero just fifteen years ago to over 50 per cent now. With just about 10 per cent of global oil production, Asia accounts for a quarter of global oil consumption. IEA predicts that this import dependence will deepen in the years to come. The gulf between demand and supply is bridged by the Persian Gulf. Over two-thirds of India's oil imports and 44 per cent of China's come from the Persian Gulf–Middle East region (Chatham House 2007). Excessive reliance on a single region poses its own threats in a post 9/11 world, where diversification of supply is considered the key to energy security. Both India and China have to negotiate the choke point of the Straits of Hormuz as they ferry their oil out of the Persian Gulf ports.

There are, therefore, many commonalities and complementarities between India and China that point to potential synergies which could lead to cooperation and even collaboration. This chapter examines potential areas for energy cooperation between the two countries and stresses the need for deliberate, planned initiatives to realize this potential.

Potential areas of cooperation

Coal

Despite the burgeoning demand for oil, coal still claims primacy in China's energy mix, accounting for as much as 67 per cent in 2007. The National Development and Reform Commission (NDRC), the agency entrusted with the planning and implementation of China's energy security policy, envisages that coal will continue to play a dominant role in China's economy even in 2010 (Table 7.1). This view is corroborated by the 'Scramble' scenario put out by Royal Dutch Shell's just released 'Energy Scenarios to 2050'. The study envisages a ramp up of coal consumption in the next few years until environmental concerns force policies and action to the contrary (Shell 2008).

In 2006, China burned 2,392 million tons of coal, accounting for 39 per cent of the world total. A substantial portion of this went into the production of electricity. Since coal is the most abundant fossil fuel domestically available, it stands to reason that China will build more coal-based thermal power plants in the next few years. According to an estimate, 70 per cent of 1,260 gigawatts of new power stations to be built in China up to 2030 will be coal-based (Chatham House 2007).

The predominance of coal in China's energy mix presents opportunities for India, which also has an energy basket heavily weighted in favour of the same fossil

Table 7.1 Share of fuels (per cent) in China's energy mix

Source	*Consumption 2005*	*Consumption 2010*
Coal	69.1	66.1
Oil	21	20.5
Natural gas	2.8	5.3
Nuclear power	0.8	0.9
Hydropower	6.2	6.8
Other renewable energy	0.1	0.4

Source: National Development and Reform Commission, Dec 2007.

fuel. Currently coal accounts for just over half of India's energy basket, most of it being used for thermal power generation. This growing reliance on coal, partly explained by the unavailability of sufficient quantities of gas and uranium and the staunch opposition to construction of big dams, is likely to continue through this decade as well as the next. India is currently licensing Ultra Mega Power Projects of 4,000 MW capacity to cope with its electricity shortages. The UMPP developers are given captive coal mines as well as a host of other incentives. With over 78,000 MW of capacity planned for the Eleventh Five Year Plan period ending in 2012, coal seems to be the quickest way to achieve the target.

In the circumstances, India is looking for efficient and cleaner ways to burn coal, including deployment of supercritical boilers and IGCC (integrated coal gasification combined cycle) technologies. China has also stated that it intends to develop 6,000 megawatt ultra-supercritical units, large IGCC units and combined heat and power technologies (NDRC 2007). China and India have a synergistic interest in efficiency-enhancing coal technologies and therefore development of coal technologies could be a potential area of cooperation between India and China.

Both India and China are also exploring possibilities to liquefy coal so that they could reduce their dependence on imported oil. India now offers captive coal mines to potential investors in coal liquefaction projects. In China, Shenhua Group Corporation Limited, one of China's largest coal producers, launched the coal liquefaction project in 2004 in Erdos, a city in the Inner Mongolia Autonomous Region. The plant will produce 6 million tonnes of liquid fuels and will commence production from 2008. South African firm Sasol, which pioneered the technology, is likely to partner projects in both India and China. Sharing of experience in this emerging technology could be mutually beneficial. Coal – its use in an efficient and environmentally sustainable manner – presents the most significant opportunity for cooperation in a wide range of activities from R&D to manufacture, construction, operation and maintenance of coal-fired plants as well as related technologies. India has also made some advances in extraction of coal-bed methane and this is another area where India and China could explore synergies.

Carbon capture and sequestration

Pollution is an ineluctable concomitant to a coal-dominant growth paradigm. Heavy reliance on coal makes China the leading emitter of air contaminants and GHGs. As early as 1995, China's sulphur dioxide (SO_2) emissions ranked the worst in the world, accounting for 24.2 per cent of the world total. Since then, China's SO_2 emissions have increased steadily. China emits about 14 per cent of global carbon dioxide and this is predicted to rise to 17 per cent by 2020 (Chatham House 2007). A preliminary estimate by the Netherlands Environment Assessment Agency reveals that China has overtaken the United States as the world's biggest producer of carbon dioxide in 2006, fuelling anxiety about China's growing role in human-induced climate change.

India, for its part, is also treading a carbon-intensive growth path, although with 17 per cent of global population, the country accounts for just 4.6 per cent of global GHG emissions. The Integrated Energy Policy, India's first comprehensive policy document drawn up by the Planning Commission in 2006, chalks out various policy options for India and settles on a coal-dominant growth paradigm as the most likely scenario. India's National Communication submitted to UNFCC estimates that GHG emissions from all activities reached 1.2 billion in 1994. However, since then the Indian economy has been growing rapidly and its emissions concomitantly. According to one estimate, in 2000, GHG emissions reached 1,484 billion tons of carbon equivalents (Sharma *et al.* 2006).

The energy choices that India and China make now are therefore critical to determining the effectiveness of the global battle against climate change. Despite the failure of the Bali conference to come up with implementable solutions to arrest global warming, it is inevitable that any scheme to contain and reverse global warming will have to necessarily involve China and India. The Clean Development Mechanism, despite its pious objectives, can hardly provide the solution for carbon-free growth in developing economies. Developing countries, especially China, Brazil, India and South Africa will have to shift to a carbon-free growth paradigm not only immediately, but also on a scale that will achieve a meaningful freeze of global GHG emissions. Acknowledging the need for access to clean technologies, China wants international technology transfer and cooperation strengthened to share the benefits of technology worldwide.

> Measures in this regard will include the following: establishing an effective technology cooperation mechanism to promote R&D; deployment and transfer of technology of addressing climate change; eliminating obstacles to technology cooperation in terms of policy, institution, procedures, financial resources and protection of intellectual property rights.
>
> (NDRC 2007)

Last year, the United States and China agreed to strengthen cooperation in advancing clean coal technology, aiming to develop up to 15 large-scale coal-mine methane capture projects in China, providing policy incentives to abolish cost barriers to full commercialization of advanced coal technologies, and advancing the research and development of carbon capture and storage technologies. Laudable as these efforts are, technology for carbon capture and sequestration should be freely available to all developing countries dependent on fossil fuel-driven growth. Joint lobbying by China and India for dissemination of clean coal and carbon sequestration technologies could break down barriers erected by the global trading regime in the form of IPRs and patents.

Oil

The year 1993 marked a watershed in China's energy frontier when, after three decades of self-sufficiency and several years of oil exports, the country became a net oil importer. Domestic oil production, especially from mature on-shore fields has been stagnating and China is now hoping to ramp up production in its western oilfields in Xinjiang. Even so, imports will have to quench China's ever-growing thirst for oil. The International Energy Agency (IEA) had estimated that by 2010, imports will rise to about 200 million tonnes a year, a figure that has already been breached. China's oil demand in 2008 could reach 400 MT per year. China has already outpaced Japan, to become the second largest energy consumer in the world. Table 7.2 provides estimates of China's oil demand by various agencies. Imports are set to go up as China consumes more oil.

With transportation driving oil demand, it is not surprising that China should have galloping growth in oil consumption. In 2001, China had 15 million cars, buses and trucks, but automobile manufacture is proceeding at a frenetic pace with 2 million vehicles being added every year. It is therefore plausible that by 2020, oil imports will reach 400 million tonnes annually[4]. At current levels of production that would mean an import dependence of over 70 per cent.

India's dependence on imports stands already at 75 per cent of domestic consumption and hence is even more acute than that of China, although, in absolute quantities, India's oil demand is much smaller. As in the case of China, a substantial part of India's growth is hydrocarbon-led. India's automotive industry, which accounts for a 3.2 per cent share in the country's GDP, has a turnover of

Table 7.2 China's oil demand (million tonnes per year)

	2000	*2004*	*2010*	*2020*	*2020 Imports*	*Import Share*
Actual (BP 2005)	5.0	6.7				
IEA (WEO 2004)			7.9	11.6	7.1	67%
DOE (IEO 2005)			9.2	12.3	8.8	72%
East-West Center (3/05)			8.6	12.3	8.5	72%
IEE Japan (3/04)			7.3	12.0	8.5	71%
Merrill Lynch (11/04)		10.0				

Source: National Bureau of Research, April 2006.

$145 billion and provides 12 million jobs. The Indian Government plans to increase this share to 8 per cent of GDP, providing 25 million jobs by 2016 (GOI 2007). There seems little doubt that without major new finds India's imports will go up steeply in the next few years.

An emergency stockpile is integral to energy security in oil-importing nations. Ever since the 1973 oil crisis, importing nations around the world have scrambled to build an emergency stockpile of crude and petroleum products – an inventory that could be used during periods of physical supply disruption. The US has an SPR of about 95 million tonnes – the largest in the world – while the EU also has a common stockpile under the auspices of the International Energy Agency. Japan and South Korea are also members of the EU stockpile. Though located in individual member territories, the management of the stockpile is a joint EU responsibility. In today's world, where terrorist threats to supply disruption are no longer in the realm of fantasy, no import-dependent country can afford not to have a strategic petroleum reserve.

Both China and India have decided to set up their respective stockpiles – of 90 days eventually in the case of China and 15 days, initially, in the case of India. In 2004, China identified four sites for its national petroleum reserves tanks and began building its strategic petroleum reserve (SPR) in three phases, to be completed by 2020. The first phase, to be completed by 2008, will hold 100 million barrels (13.6 million tonnes) – equivalent to 25 days of China's net oil imports. The second phase is planned to add 200 million barrels, covering 42 days of net oil imports. The completion of the third phase may increase the net storage capacity to 500 million barrels after 2010 (Harbert 2007). The high oil prices of recent years prompted China to delay oil purchases to fill its strategic reserve. Only one of the four sites in Phase I, consisting of 52 storage tanks – a total capacity of 33 million barrels (4.5 million tonnes) – has been filled. In addition to the national SPR, the government will reportedly require major state-owned oil companies to hold government-mandated oil stocks. China's SPR regulations are part of a comprehensive Energy Law (Harbert 2007).

India has also identified the sites for its strategic stockpile and has set up a company which would oversee its establishment and management. India and China can cooperate in the construction and joint management of crude stockpiles. Joint bidding for construction could bring down the costs. The IEA has invited both India and China to join the IEA's collective emergency response system. The details of participation have not yet been worked out, but without the inclusion of the two Asian majors, the IEA can be only partially successful in stabilizing oil markets. Meanwhile, India and China can prepare a coordinated response to the invitation to join the IEA. This is a synergetic area where there is scope and need for cooperation.

All Asian countries pay a premium of USD1 to USD1.50 for each barrel of oil imported from the Persian Gulf, thanks to a historical accident of pricing. Thus all Asian consumers pay, over and above the high crude prices, a premium which costs them collectively over US$10 billion annually. The complementarities of Asian economies call for concerted action. While there has been some discussion among

Asian buyers of the need to collectively bargain for elimination of the 'Asian premium', no concrete measures have emerged so far. Besides, proposals for an Asian marker crude – to reflect the heavier crudes consumed by Asia – have not taken off. China and India could push for an overhaul of the international oil pricing mechanism to a less discriminatory and more equitable one. Mutual support for each other's position for the development of an Asian marker crude reflecting the heavier and high-sulphur crudes consumed by Asia is another area of multilateral cooperation in which China and India can take the lead.

Natural gas

Natural gas, which accounts for nearly a quarter of the energy mix of developed countries, plays only a marginal role in China though in India it makes up a higher share, about eight per cent of the energy mix. It is inevitable that both India and China will see substantial increases in gas consumption in the coming years, especially in a climate-stressed world in search of cleaner fossil fuels. Exxon Mobil, the operator of Sakhalin 1, is in talks with China to supply piped gas from the Sakhalin 1 project. An agreement on gas price is yet to be reached. China has also entered into an agreement with Turkmenistan to build a pipeline to its western provinces. China's West-East pipeline, the longest gas pipeline system in the world, will carry gas from the country's western provinces to the markets in Shanghai.

India has substantial domestic gas production which will soon be augmented by production from new offshore fields discovered recently by Reliance, a private sector conglomerate with extensive investments in the entire petroleum value chain. As well, India is pursuing neighbourhood pipelines vigorously, the Iran–Pakistan–India pipeline talks having made substantial progress in recent months, after the visit of President Ahmedinijad to New Delhi in April 2008.

While pipelines will supply some quantities of gas to both the countries, LNG is also likely to play an increasingly important role, as liquefaction costs are driven down by constant technological upgrading. However, the potential for LNG to supplement piped gas will depend crucially on the price factor. Globally, LNG price is linked to crude price – an individual marker or a basket of crudes. In the current scenario of spiralling crude prices, such linkage may render LNG beyond the purchasing capacity of Asian markets, especially China and India. In fact, many planned LNG terminals failed to materialize in India and China as gas prices floated up and away along with soaring crude prices. Therefore, if gas suppliers could be persuaded to delink the LNG price from crude, they would find ready markets in Asia. Commonality of purpose will drive this cooperation. It will be an uphill task, but perhaps not an insurmountable one, especially since gas prices linked to today's crude prices will find few takers. Gas being a clean fuel, the support of countries concerned over climate change can be harnessed to persuade gas producers. Gas prices can be linked alternatively, to coal or fuel oils which gas replaces. If successful, it could set a new benchmark and enable gas to realize its potential as the fuel of this century.

Electricity

Transnational electricity trade is another area of potential trilateral cooperation. Tajikistan and Kyrgyzstan, the two Central Asian Republics, have substantial hydel potential. Tajikistan has a potential of 263.5 billion kilowatt hours – of which only six per cent has been exploited. Kyrgyzstan has 163 billion kilowatt hours of potential, of which only 10 per cent has been exploited. The potential in these two countries remains unexploited for want of capital as well as markets.

If new hydel projects can be built in these countries, electricity can be transported to India through the Xinjiang district of China. HVDC lines have now made it possible to transmit electricity over long distances with minimal line losses. The Xinjiang route will make it possible for India to by-pass Pakistan and reach Himachal or Ladakh in India. No doubt, the lines will have to traverse difficult mountain terrain, but Powergrid Corporation of India – the country's state-owned transmission company – has demonstrated expertise in building high-voltage transmission networks in difficult terrain.

During the Soviet era, all the Central Asian Republics were interconnected through a high voltage grid and there was power flow between Russia and the Republics. In recent times, however, the grid has been disconnected and is in disuse, although both Kyrgyzstan and Tajikistan supply some electricity to Uzbekistan through cross-border connectivity. It might be worthwhile to explore whether the Soviet period grid connectivity cannot be restored at minimal cost. In fact, in the future, it is also possible to extend the connectivity to Uzbekistan where electricity generated by gas turbines can be fed into the same grid to be transported southwards to India.

China and India can also collaborate in the construction of hydel projects as well as transmission networks, pooling capital as well as technology. For India, these projects will have the added advantage of bypassing troubled Afghan and Pakistani territories. Additionally, China can earn transit revenues as well as help in the development of Xinjiang. India's new electricity law allows open access to the country's transmission and distribution networks so that today bulk consumers can directly buy power from any producer.

Large-scale cross-border power trading is not a novel concept. In fact, it has been successfully practiced in many parts of the world, such as Canada – United States, United States – Mexico, Scandinavia – Germany, and France – United Kingdom. Islanding technologies now make it possible to isolate problem stretches of the grid.

While the relative economics of a gas pipeline vis-à-vis HVDC cable need to be determined, in the event that the latter turns out to be cost effective more prospects open up. It might even be possible to eventually transmit electricity all the way from Xinjiang – fired by its gas reserves – to India, tapping along the way, Kyrgyzstan's and Tajikistan's hydel potential. Considering this potential, the time has come to seriously study the possibility of setting up a trans-Asian electricity grid on the lines of the EU electric grid.

Renewables

China acknowledged the importance of renewable energy a decade ago when its State Development Planning Commission and Ministry of Science and Technology launched a joint Incentive Policy for Renewable Energy. In its Tenth Five Year Plan document, renewable energy became the poster-child of the power sector. Photovoltaics and wind and geothermal technologies became thrust areas. In a far-reaching initiative taken in 2005, China enacted a new Renewable Energy Law effective from January 2006. The law targets 15 per cent electric supply from renewable sources and mandates feed-in tariff for electricity from renewable sources. NDRC will also implement a national renewable energy plan, including specific renewable energy targets that will act as the framework for implementation of the law. Provincial planning agencies will then develop their more specific implementation plans. Arming the agency with teeth, the law provides for specific penalties for non-compliance with its provisions.

Within the renewable segment, China intends to increase wind-power capacity to 1.2 gigawatts. A Greenpeace study finds that China's Guangdong province alone has the potential for 20 gigawatts of wind capacity by 2020 (Greenpeace 2005). Total wind potential in the country could be as high as 3,200 gigawatts. In 2007, China's installed wind capacity rose to 6,000 megawatts as against India's 8,000 megawatts, a substantial part of it added in 2007. India has made great strides in wind energy in the last few years and now has the fourth largest wind capacity in the world. Already Suzlon, an Indian wind energy major that has carved up a fifth of the global wind turbine market, has set up shop in China. Suzlon manufactures and supplies wind turbines which now account for eight per cent of all wind turbines in China. With huge domestic potential in both countries, this is an area ripe for collaborative R&D to bring down the cost of wind turbines. Similarly, collaborative R&D in solar photovoltaics and other renewable sources would be mutually beneficial. Decarbonising transport could be another thrust area for collaborative research.

Others

Energy efficiency and conservation are other areas where India and China could reap enormous benefits through a cooperative approach. There is scope for China and India to reduce the energy intensity of their economies through the use of more energy-efficient appliances. Cross-investments in each other's energy sector are a potentially viable area of cooperation. Both countries have opened up their energy sector to overseas investors, although to varying degrees. China is planning to acquire upstream acreages for exploration in India through the NELP (New Exploration and Licensing Policy) licensing rounds. Chinese companies have also successfully bid for the construction of a pipeline in India. India has demonstrated strengths in building versatile refineries and Indian investors are exploring overseas investment opportunities. Indian industry could explore the possibility of setting up refineries in China. China and India have also cooperated intermittently in their quest for overseas oil acreages. They can go it alone where it suits their interests, but can also benefit from cooperation where opportunities exist. However, both countries could agree to avoid competing with each other – a process that only goes to push up the price of the oil asset without any concomitant benefit to either country.

Finally, India, the United States and China have come together in an alliance of major oil importers with the objective of cooperating with each other to stabilize global oil prices. Just as OPEC has been a successful producers' cartel, the time has come for a global consumers' cartel so that predatory pricing practices are dealt with firmly and effectively.

Conclusions

This chapter has identified some broad areas for cooperation between China and India, but the list is not exhaustive. The areas span the spectrum of fuels, geo-political issues and market related interventions. There could be areas of conflict and competition which foreclose possibilities for a cooperative approach, but there are enough synergetic areas where cooperation is mutually beneficial.

However, effective cooperation will require both countries to initiate a whole host of measures in sequence. A top level meeting between leaders of both countries is important to agree upon broad areas of cooperation in the energy sector. Joint working groups of policy makers and academics could then identify specific areas where cooperation would be mutually beneficial and examine the policy and legislative changes needed to proceed further. Consultations with scientists and industry representatives from both countries would then help to take specific proposals further and demarcate the framework for cooperation. In order to sustain cooperation (especially between countries such as China and India that have a history of competition and conflict), identifying specifics and ensuring follow-up are as vital as formulating ideas and articulating a vision. Moving from the general to the specific is key to transforming rhetoric into reality.

Notes

1 This chapter is a revised version of the paper presented at the TERI-KAF Conference on 'India's Energy Security: foreign, trade and security policy contexts', 29–30 September, 2006.
2 US imported goods worth $321 billion from China in 2007: http://www.census.gov/foreign-trade/balance/c5700.html#2008 (accessed on May 21, 2008).
3 According to the Ministry of Power, more than half the rural households in India do not even have connectivity to the electric grid.
4 IEA estimates are higher than those of others. Baker Institute: 3 mbd, APERC: 2.9 while Chinese estimates are much lower. PRC State Council: 1.7.

References

Chatham House, 2007. *Changing climates: interdependencies on energy and climate security for China and Europe*. Chatham House Report. London: Royal Institute of International Affairs.
GOI, 2007. *Automotive missionpPlan 2006–2016*. Dept of Heavy Industries, Govt of India.

Greenpeace, 2005. *Wind Guangdong*. Greenpeace, Sun Yat-sen University, Guangzhou, China.

Harbert, K. A., 2007. *China's energy consumption and opportunities for US-China co-operation to address the effects of China's energy use*. Presentation to the US-China Economic and Security Review Commission, June 2007.

IEA, 2007. *World energy outlook 2007*. Paris, France: International Energy Agency.

NDRC, 2007. *China's national climate change programme*. National Development and Reform Commission, People's Republic of China.

Sharma, S., Bhattacharya, S., and Garg, A. 2006. 'Greenhouse Gas Emissions from India: A perspective'. *Current Science*. Special Section: Climate Change and India. 90(3). pp 326–333

Shell, 2008. 'Shell energy scenarios to 2050.' Shell International BV, viewed 15 May, 2008 <http://www-static.shell.com/static/aboutshell/downloads/our_strategy/shell_global_scenarios/shell_energy_scenarios_2050_2008.pdf>.

8 Security of maritime energy lifelines

Policy imperatives for India[1]

Commander G. S. Khurana, Indian Navy

Introduction

'It is estimated that … the world will become tripolar by about 2050–55 with China, USA and India at the high table. China and India that once contributed to half the global wealth will return to this position after a quarter of a millennium and the global strategic centre of gravity will be back in Asia' (Bhaskar 2006).

The economic (and developmental) index of a state is closely linked to its need for energy, and this explains India's quest for a larger share of the world's resources. Its emergence as a major player in overseas ventures is already beginning to have a palpable effect on global geopolitics, particularly when it 'stretches' the global oil supply and adds on to the 'China-effect'.

India's energy-demand profile

India's need for energy resources is primarily rooted in imperatives triggered by its accelerated development. The country's current commercial energy consumption is a mere 25 per cent of the world average – an apt indicator of the potential to be realised in the years to come. Furthermore, as much as 73 per cent of India's industrial energy needs are currently met by coal, which is being replaced by natural gas and petroleum products (Ghosh 2006). As the country's manufacturing sector develops further, the imperatives of power generation will lead to an increase in the demand for natural gas. Despite new gas finds since 2002, India's own resources are inadequate for her burgeoning needs.

The demand for oil resources is also rising rapidly due to the demands of public and freight transportation (sea, air and land) and the increasing number of private vehicles on Indian roads. During the past decade, India's crude oil consumption has grown by over 6 per cent annually, which is twice the world average growth (Tønnesson and Kolås 2006). While its annual oil demand was 40 million tons a decade ago, it is now 135 million tons and is expected to increase to 370 million tons in the next two decades (Aziz 2006). However, the domestic production of crude oil has been stagnant and the current oil-import dependence of 70 per cent is likely to increase to 90 per cent by 2030 (IEA 2005). This explains the critical nature of external supplies of energy resources for India.

Insular India

India's crude-oil imports are all brought in by sea and, for many years to come external supplies of natural gas may also be transported as Liquefied Natural Gas (LNG) via the maritime route. The options for overland natural gas pipelines to India have been few. The projects mooted so far include the *Oman–India* (submarine) pipeline, the Turkmenistan–Afghanistan–Pakistan–India (TAPI) pipeline, the *Iran–Pakistan–India* (IPI) pipeline, the *Bangladesh–India* pipeline and the *Myanmar–Bangladesh–India* pipeline. These proposals have either been found unfeasible (technically or economically), or have often run into 'rough seas'.

In terms of overland communications, India is bottled up within the sub-continent, which incidentally, also bears its name. This has led to 97 per cent of India's trade (by volume) being by sea, which is comparable to that of an island state. The barriers are posed by two important factors – the first is natural topography (viz. the highlands stretching across India's northern frontier) and the second is political discord with Pakistan and Bangladesh. Russia and the Central Asian Republics (CAR) have emerged as the new 'petro-states' with ample natural gas reserves. However, owing to the aforesaid barriers, their reserves are inaccessible to India despite their geographical proximity. The various problems that have almost completely stalled progress on the Iran–Pakistan–India gas pipeline project typify the barriers to overland trade. In the east, Bangladesh has declined to provide its own surplus gas to India, while the proposal for a transit corridor for Myanmar's gas is mired in political wrangling between India and Bangladesh. A pipeline from Myanmar through India's north-east, (bypassing Bangladesh) is being considered, but this too involves an 'inhospitable' terrain and insecurity due to ongoing insurgencies in the north-eastern states. Besides, it is doubtful if Myanmar would have sufficient gas for India (after providing it to China), for a pipeline to be cost-effective. It thus remains to be seen whether any of the proposed gas pipelines will ever reach India, but even if they materialize in the long term, most natural gas imports would still continue to arrive by sea.

In the case of domestic oil and natural gas, two-thirds of domestic production is presently sourced from offshore locations in India's maritime zones. When the United Nations Convention on the Laws of the Sea, 1982 (UNCLOS-3) came into force in November 1994, it provided for a *sui generis* regime of an Exclusive Economic Zone (EEZ) for all coastal states, extending to a distance of 200 nautical miles (nm) from the coast.[2] Given the depletion of natural resources on land, this was a boon for all these states, and India was no exception. In fact, considering its 7,500 kilometre-long coastline, which is the fifteenth longest in the world, India was given sovereign rights over an extensive maritime zone of 2.2 million sq km, which is nearly two-thirds of its total land area. India's widespread island chains also contributed substantially to this large EEZ. The maritime zone around the Andaman and Nicobar Islands, for example, makes up 30 per cent of India's total EEZ. Besides other resources, the seas under India's jurisdiction have provided some much-needed oil and natural gas resources to India; with a promise for more. By May 2009, India is required to submit its claim for an additional maritime zone, called the Legal Continental Shelf (LCS), which would extend beyond the EEZ up to 350 nm from the coast (or less, depending upon the underwater topography) and provide an additional seabed area of about 1.5 million sq km.

Aim of the chapter

Ensuring good order in India's maritime zones and the safe and unimpeded passage of shipping responsible for transporting fossil-fuel resources through the Sea Lines of Communication (SLOC), thus constitute major components of India's energy security. However, the overall security scenario has become more complex over the years, particularly in the Indo-Pacific region, carrying the potential to severely affect national security and retard India's economic growth and development. The 'good news' is that India's growing comprehensive power and its geo-strategic location astride major sea routes translate into greater capacity for the Indian state and enable its growing maritime (including naval) power to respond to these insecurities. These factors also open fresh avenues for engagement and security co-operation with regional states and extra-regional powers. This chapter aims to analyse these issues in the context of India's geo-strategic imperatives and to deduce an appropriate policy response towards ensuring secure 'maritime logistics' of energy resources.

Expanding 'stakes'

As noted earlier, offshore locations in India's EEZ account for two-thirds of India's domestic production of oil and gas. They include Bombay High off the western coast and the various river basins off the eastern coast, such as the Krishna-Godavari, Cauvery and Mahanadi basins. New energy finds in Indian waters are also expected in the years to come. The hydrocarbon potential of these offshore areas is high considering the 'encouraging' response of oil and gas companies so far to the exploration acreages offered by the government. The results of the sixth round of auctions were declared in mid-September 2006. Fifty-two bids were received for 24 deepwater blocks and 25 bids for six shallow offshore blocks. Beside the Indian companies like the Oil and Natural Gas Commission (ONGC), Reliance Industries, Reliance Natural Resources and Oil India, the bidders included many foreign firms, either in partnership with Indian companies or alone. They included British Petroleum, Naftogaz of Ukraine, PGNIG of Poland, Petronas of Malaysia and Cairn Energy of the UK. Most of the offshore blocks are in the Bay of Bengal.[3]

The security of offshore areas and extraction platforms, widely dispersed over 40,000 sq km of maritime area, is of vital importance to India. Presently, extracted oil and natural gas resources are mainly transported to refineries and consumers ashore via submarine pipelines, since the distances involved are relatively short (the longest submarine pipeline is about 200 kilometres long); and tankers are primarily used for the coastal movement of refined products. While submarine pipelines are inherently secure relative to tankers, India may need to rely more on tankers in the future. The offshore areas off Andaman and Nicobar Islands are said to possess the largest reserves of coal bed methane (CBM) in the country (Choudhury 1998). The ONGC has also identified prospective areas in the offshore Andaman and the Laccadive ridge in the Arabian Sea containing natural gas hydrates. Hydrocarbon reserves may even be located in the legal continental shelf beyond the EEZ. This implies that, in years to come, offshore resources and products are likely to be transported to the mainland over significantly large transit distances, thereby necessitating the use of tankers/shipping (since laying submarine pipelines may not be cost effective).

In the case of India's imports of fossil-fuel resources, more than 60 per cent of its crude oil is presently sourced from the West Asian (Middle East) region. While Saudi Arabia is India's largest supplier, the rest of the oil comes from other countries like Iran, Kuwait, Oman and the UAE. Some African states have also become India's major crude oil suppliers. Nigeria is currently India's second largest source of crude oil. A long-term contract for LNG presently exists only with Qatar, but similar agreements are likely to be finalized soon with other countries like Algeria[4]. India has also been resorting to spot buys of LNG, such as from Algeria and Egypt.

India's ONGC Videsh Ltd (OVL) or other companies like the Indian Oil Corporation (IOC) and the Gas Authority of India, Ltd (GAIL) have also acquired energy stakes in many countries overseas, which have the potential to meet future energy needs. Iran, Iraq and Syria are among these in West Asia. In Africa, OVL has acquired stakes in Sudan, Libya, Egypt and Cote d'Ivoire (Ivory Coast).

India initially turned to Africa to reduce its heavy dependence on West Asia, but it has also been making efforts towards source diversification to the East. Indian oil and gas companies hold a 45 per cent stake in Vietnam's gas fields, a majority (55 per cent) stake in Australia and a 20 per cent stake in Myanmar's gas fields (all offshore). In case the Myanmar–India gas pipeline proposal seems unfeasible, India could resort to transporting the gas as LNG. A Malaysian Petronas official has indicated that a long-term contract is being negotiated by India to import LNG from Malaysia.[5] In the Russian Far East, OVL acquired a 20 per cent stake in the Exxon Mobil-operated Sakhalin-I oil project. The first two crude oil consignments of 700,000 barrels each were to be shipped to India in October and December 2006.[6] Plans are also afoot for OVL to export its share of natural gas from Sakhalin-I as LNG.[7] The imperative to diversify energy sources has also compelled India to look as far away as the Americas. OVL had earlier acquired 30 per cent stakes in six deep-water exploration blocks in Cuba's EEZ. In September 2006, it signed a production sharing agreement with Cuba's national oil company Cupet (Cuba Petroleos) for an additional two blocks. Also, ONGC and GAIL have jointly taken up a 49 per cent stake in a Venezuelan oil field (Zora and Woreck 2005). Crude oil supplies from Venezuela have since begun to arrive in India.

Coal will continue to constitute a major proportion of India's energy mix for many years to come. Although India ranks as the third-largest coal producer in the world next only to China and the USA, the shortfall between demand and domestic production is expected to be 105 million tons by 2011–12.[8] Even in the past, some coal had to be imported from Australia and New Zealand since India's own resources are of low quality, with high ash content.

It is well known that, over the years, India's strategic interests have significantly expanded overseas, westwards and eastwards in terms of India's overall trade interactions with the world. This is true for its energy security as well. While geographical proximity to West Asia is an advantage due to the lower costs for oil and gas transportation, India is at a significant geo-strategic disadvantage when it comes to diversifying energy sources, since it is straddled between maritime choke points – the Straits of Hormuz/Bab el Mandeb passage to the west; and the Southeast Asian straits to the east. In any case, West Asia will almost certainly continue to be India's principal energy source. As former Ambassador Ishrat Aziz put it in a recent article on India's strategic imperatives: 'while the (Persian) Gulf oil will last for 80 years, the non-Gulf oil will severely deplete in 15 years, unless production patterns change' (Aziz 2006).

It is important to note that most of the countries from which India's energy resources are sourced or where India has acquired energy stakes are plagued by geopolitical and social instabilities. In most cases this has a maritime-security dimension since oil and gas fields are sited offshore. However, the inherent political and financial risks in opting for these overseas investments were inescapable. Of course, India has been building close political ties with these resource-rich countries since energy concerns have become an important element of foreign policy, but they may not suffice as insurance for its energy security.

Maritime insecurities

Non-state threats

The predominant insecurity to sea lines today is brought about by non-state actors. Piracy and armed robbery of ships, for which the Southeast Asian straits have been infamous since historic times, is the foremost among these. Since 2005, the menace has become rampant off East Africa too, particularly off the coast of Somalia and in the Gulf of Aden[9] and is now associated with better planning, precise coordination and greater violence. As the examination of India's expanding energy stakes (in the previous section) indicates, much of Indian shipping that is carrying energy resources has either commenced traversing these areas, or will do so shortly. There have also been instances in the past when piracy has spilled over from the Southeast Asian waters westwards into the Bay of Bengal or further east into the South China Sea. At times, even some insurgent groups such as the Free Aceh Movement (GAM) in Indonesia and the Liberation Tigers of Tamil Elam (LTTE) in Sri Lanka have resorted to hijacking commercial vessels to further their political agenda. While the threat from GAM has receded lately, after its cease-fire with the Indonesian government, the LTTE threat to the 'arterial' shipping route transiting south of Sri Lanka cannot yet be discounted, even though the separatist group has a confined political agenda. In August 1998, the Belize-flagged ship *Princess Kash* was hijacked by the LTTE for the cement that it was carrying, which the Tamil Tigers needed to build bunkers, tunnels and other underground infrastructure. The Sri Lanka Air Force bombed it to prevent the LTTE from seizing the cargo (Gunaratna 1998). The LTTE has also been striking oil targets at sea. In October 2001, it carried out a coordinated suicide attack by five boats on the oil tanker MT *Silk Pride* off northern Sri Lanka (Luft and Korin 2004).

In addition to piracy, the probability of widespread terrorist attacks on shipping by Islamic radicals has increased substantially in the past few years. Al Qaeda is firmly entrenched in West Asia and its rudimentary nautical expertise, targeting stationary ships using fast boats packed with explosives, was demonstrated with the USS *Cole* attack in October 2000. Using the same modus operandi, the slow-moving French supertanker *Limburg* was attacked off Aden in October 2002. The attack was accompanied with Bin Laden's ominous warning to the West, 'By God, the youths of God are preparing for you things that would fill your hearts with terror and target your *economic lifeline* until you stop your oppression and aggression.'[10] Soon thereafter, energy infrastructure became the key targets of global terrorism, as indicated by events in the Persian Gulf, including the April 2004 Al Qaeda suicide attacks on Iraqi Al Basra and Khor Al Amaya offshore oil terminals.[11] Notwithstanding the presence of multinational naval forces, the Persian Gulf region continues to remain unstable. Besides instigation by Al Qaeda, the violence is closely linked to instability in Iraq, which is nowhere close to normalcy as yet. Many incidents of piracy have also been reported in the northern Gulf. The security scenario could deteriorate further when the coalition naval presence is eventually scaled down. With 90 per cent of oil exports from the Persian Gulf being carried by sea (see EIA 2007) the threat to the Gulf's offshore platforms, loading berths and oil/ LNG tankers remains pronounced. This has also significantly enhanced the risk to shipping transiting choke points like the Strait of Hormuz and Bab-el-Mandeb.

Al Qaeda's offshoot, Jammah Islamiyah, is active in Southeast Asia, where it has been fuelling separatist movements with jehadi calls to insurgents operating in various states like Indonesia, the Philippines and southern Thailand. The Jammah Islamiyah aims to establish a pan-Islamic state in Southeast Asia. The 'insecurity-triad' in the region, formed by maritime crimes (particularly piracy), separatist movements and Islamic fundamentalism, creates clear concerns regarding the shipping transport of energy supplies through the regional straits. In particular, the Malacca (and Singapore) Strait(s) is the most threatened. It is the busiest of all, through which a quarter of the global trade passes aboard more than 62,000 vessels every year, including half the world's oil and two-third of its LNG. It is widely feared that terrorists may use the 'nautical skills' of the pirates to carry out attacks on shipping in the Straits; after all, boarding ships is the 'livelihood' of pirates. Although there has been no concrete evidence yet of such a nexus, the threat remains real. Terrorists may either mimic the pirates or even employ them to hijack ships. Another possibility goes further; some among the jehadis may be 'super-specialized' in piracy as a source of generating funds, considering ongoing international efforts to freeze their finances. It is thus not surprising that some incidents of piracy/hijacking in the Southeast Asian straits are being considered as precursors to widespread maritime terrorism. In February 2004, the Filipino group Abu Sayyaf bombed a passenger ship, *SuperFerry 14*, killing 116 people (Cruz 2004). In March 2003, the hijacked Indonesian tanker *Dewi Madrim* was steered by pirates in the Malacca Straits for one hour before the criminals fled with technical documents. This case is being widely considered as a probable preparation for a maritime version of 9/11 (Elegant 2004), since the control of a vessel would provide the terrorists with innumerable attack options. It may not be practically possible for terrorists to capsize a large vessel at the most critical spot to block the Malacca Straits (or even the Hormuz Strait), but a major attack on a hub-port, a cruise liner or use of a radiological weapon in the area would create sufficient psychological reverberations to effectively and severely disrupt global energy supplies. It is also feared that the terrorists could use one of the large vessels transporting LNG as a 'floating bomb' to destroy a hub-port. Studies have, however, indicated that this fear is unfounded since natural gas is inert during transportation as a liquid. In case the tank is breached, the LNG would vaporize into its inflammable gaseous form in contact with the warm sea. But it will also slowly dissipate into the atmosphere and if ignited, the gas will not explode. An LNG carrier is therefore an unattractive option for terrorists, relative to vessels loaded with other dangerous cargo such as ammonium nitrate. Nevertheless, attacks could be carried out on LNG carriers to disrupt energy trade.

The number of tankers (carrying both oil and LNG) transiting the Straits of Malacca and Hormuz is expected to grow significantly in the next few decades. This would increase the vulnerability of these critical choke points not only to pirate and terrorist attacks, but also to marine accidents. It would also increase the 'eligibility' of these waterways for terrorists to use weapons of mass destruction.

While the energy transportation of the United States and its allies may be at a greater risk due to terrorism from Islamic radicals, it cannot be 'business as usual' for fast-developing countries like India, whose dependence on energy imports is surging exponentially. Any disruption of global energy supplies would seriously impinge upon India's energy security. In any case, the Indian Government has often expressed its conviction that 'the "jihad" that targets India is the same that targets the west, that Jaish-e-Mohammed and Lashkar-e-Taiba are synonyms of Al Qaeda and Taliban'. This point of view has been reiterated by the Indian foreign secretary Shyam Saran (see Bagchi 2006). Hence, the possibility of terrorists targeting India's energy lifelines cannot be discounted. Slow-moving India-bound tankers could be attractive targets for their bomb-laden boats. Further, as the April 2004 Al Qaeda strikes on Iraqi oil terminals indicated, it would be even easier for terrorists to attack fixed offshore installations in the Indian EEZ such as oil-extraction platforms and single-buoy moorings (SBM) using fast boasts laden with explosives. The offshore rigs could be attacked by scuba divers, as revealed by the seizure of a 'terrorist naval manual' following the arrest of a top Al Qaeda man, Al-Nashiri, in November 2002. The manual is a compilation of tactical tips for carrying out maritime attacks and mentions the use of underwater scooters and limpet mines to destroy fixed installations. This revelation also underscores Al Qaeda's link with Sri Lanka's LTTE. While Al Qaeda's frequently used bomb-ridden-boat tactics are well known to be motivated by the LTTE[12], the Tamil Tigers are also well versed in scuba diving.

Any maritime crimes in Indian waters contribute to 'disorder' at sea and thereby increase the risk factor for India's offshore rigs and the vessels that support fuel extraction operations. Although Indian waters have witnessed a reduction in incidents of piracy since 2003[13], the threat remains. Besides, other organized crimes like drug trafficking, gun-running and human smuggling persist. These drive internal instabilities and terrorism in Southeast Asian states, Sri Lanka and India, and thus indirectly impinge upon the security of sea lines being used to transport energy resources.

Military threats

It is often said that in present times, military threats to sea lines have receded in the wake of the economic boom and increased emphasis by states on national development, particularly in the Indo-Pacific region. The reduced focus of 'territoriality' in inter-state relations may have also contributed to this. However, the potential for military threats at sea cannot be discounted on account of exceptions like the unresolved Taiwan issue. China has kept its options open regarding the use of force for Taiwan's reunification, and if it resorts to using that option, shipping transiting through the sea lines of the western Pacific would be seriously threatened. Conflicts could also be triggered by inter-state contentions over maritime claims, which abound in the South China Sea. In the Persian Gulf too, insecurity of energy lifelines transiting the Strait of Hormuz persists due to the possibility of a military standoff between Iran and the United States over the former's nuclear programme. It is thus not surprising that states have continued to factor the control of maritime choke-points into their military calculi. Iran is known to have made contingency plans to block Hormuz, and it possesses adequate military capabilities in terms of submarines, mines and anti-ship missiles to do so.

India's political relationship with Pakistan continues to bear an adversarial character. Pakistan's lack of conventional military superiority vis-à-vis India has led to its naval doctrine laying a greater emphasis on a sea-denial strategy though the use of its missile-armed long-range patrol aircraft and modern submarines, which could be used to interdict India's strategic energy supplies in case of hostilities breaking out. In the medium term, China is also expected to establish its naval presence in the Indian Ocean. Notwithstanding the fact that Chinese motivations to do so are primarily dictated by the imperatives of energy security and that Sino-Indian political and economic ties are growing, the relationship between the two also continues to bear an adversarial potential on many counts, including the outstanding border dispute. The naval presence of China in Indian waters would further complicate India's security calculus, including its supplies of highly strategic energy resources. It would be easier for China to interdict India-bound tankers than it would be for India to choke China's energy supplies transiting the Indian Ocean. The reason is that China could declare exclusion zones in areas closer to Indian ports less frequented by 'neutral' shipping, but all Chinese energy supplies would be routed through the international sea-routes in the Indian Ocean, where it would be extremely difficult to identify the 'neutral' vessels.

Even in case of a conflict not involving India, such as over Taiwan in the western Pacific or due to a US–Iran conflict in the Persian Gulf, Indian trade could be seriously threatened, since the belligerents would be likely to declare exclusion zones in the maritime areas affected by the conflict. Even though India would be a 'neutral' state, a vessel carrying Indian cargo could be attacked, since discerning neutral shipping in the dense shipping of a sea line is extremely difficult.

The vulnerabilities

India's crude oil consumption stood at over 124.5 million tonnes in 2005 (BP 2006). With its domestic production accounting for only about 39 million tonnes, India needs a VLCC-full load each day (a typical Very Large Crude Carrier of three lakh tons dead-weight tonnage can carry about two million tons of oil). On an average, about 40 shipments (including those carrying petroleum products) arrive in India each month. Considering the growth rate of six per cent, the daily demand would double to five million barrels by 2020, and considering the stagnant domestic production this means a daily requirement of two VLCC-equivalent loads of crude oil in 15 years. Added to these shipments are increasing shipments of LNG.

This situation translates to a number of vulnerabilities for India. First, most of the vessels bring resources from West Asian and African states, where energy infrastructure (for both production and distribution) and transportation are at high risk due to the prevailing Islamic fundamentalism, geopolitical instability or ethnic tensions.

Second, the probability of a pirate attack or any other mishap occurring on a vessel carrying Indian energy supplies has increased in direct proportion to the growing number of India-bound vessels. These vessels also have to traverse pirate-infested areas, reaching India either through the Gulf of Aden/Somali waters (from the west), or through the straits of Southeast Asia (from the east). With the increasing size of tankers, driven by motives of economic profitability, the vulnerability of these vessels has increased significantly over the years. Terror strikes and boarding by pirates have become relatively easy, since the large tankers are now more cumbersome to manoeuvre and need to reduce speed while transiting restricted waters (due to depth constraints). These vessels (like other merchant ships) have only high-pressure fire-hoses for self defence. Besides, tankers cannot employ some of the latest technological advancements such as the 'Secure-ship' (a 9,000-volt electric fence installed around the ship to deter boarding by miscreants), due to the nature of their inflammable cargos. It is therefore not surprising that the 4,000-odd tankers plying the world's oceans are victims in a quarter of pirate attacks, even though tankers constitute a mere 10 per cent of the world shipping fleet.[14]

The sea line extending from the Arabian Peninsula to India also passes close to the coast of Pakistan, which makes vessels carrying India's strategic cargo extremely vulnerable to interdiction by Pakistani naval units during wartime. In a few years, when a Chinese naval presence is established in the Indian Ocean, this vulnerability would increase further. China is providing financial and technical assistance to Pakistan to develop its Gwadar port in Balouchistan Province, sited very close to the Strait of Hormuz. The facilities of the Pakistani port are likely to be used by Chinese naval units.

The increasing number of India-bound oil and LNG tankers (combined with the effect of India's growing sea-borne trade in other commodities) is also leading to congestion of shipping traffic close to Indian ports, harbours and anchorages. Collisions and oil-spills in restricted waters in approaches to harbours and ports can severely impede shipping activity and thus have the potential to disrupt energy supplies to India closer to home.

The declining share of India's overseas trade carried on its own vessels is also a matter of concern. The Indian fleet tonnage has been stagnant for many years. Although the target for India's Ninth Five Year Plan (1997–2002) was nine million gross registered tonnage (Chaudhury 1998), it could cross only the 7.6 million mark by 2006 (Mitra 2004). In terms of India's overall overseas trade, Indian-flagged vessels are not being able to keep pace with the country's growing needs, which becomes more critical when strategic imports like energy resources are involved. In 1999–2000, oil tankers comprised 2.71 million gross registered tonnage (GRT) within the total Indian shipping of 6.93 million, or about 39 per cent (UNCTAD 2000). In 2005–06, they constituted 4.45 million GRT within the total of 7.52 million GRT, or 59 per cent (UNCTAD 2005). Thus, there was a 20 per cent increase in the proportion of oil tankers over other vessels in those five years. This includes the acquisition of two VLCCs by the Shipping Corporation of India (SCI) in 2005, which amounted to a significant increase in the capacity of Indian-flagged tankers. However, considering that some of the vessels need to be laid down for maintenance and repairs, the overall capacity remains grossly inadequate. It is also important to note that more than 30 per cent of the Indian fleet is over 20 years old and needing replacement within the next five years (Mitra 2004) 25 years is the internationally accepted cut-off period for the economic life of a vessel. Many countries are imposing a ban on ageing tankers from even entering or transiting their waters.

Consequently, more than 60 per cent of the vessels chartered to carry crucial energy resources or petro-products to India fly foreign flags. There is total dependence on foreign vessels for the transportation of natural gas, given that India does not have any LNG tankers of its own. Many of these foreign vessels actually fly 'flags of convenience'. In other words, they are ships owned in one country, but registered elsewhere in countries like Panama, Honduras and Liberia (Pan-Ho-Lib countries). The practice has been encouraged in the past by some states to promote trade and the concept has almost become a norm in today's commercialised world. It is beneficial to the ship owner since the taxes are minimal. For the registering authorities, it is a means to generate revenue. More than half of the global merchant vessels today fly 'flags of convenience'. However, this increases the inherent vulnerability of the shipping industry since these vessels are often associated with non-stringent safety standards and lax manning regulations. It is virtually impossible to verify the authenticity of enlisted crew, which constitutes a potential threat, especially if one considers that, driven by economics and technological improvements, shipping companies are now employing smaller crews. Furthermore, Indonesia and the Philippines are the largest suppliers of shipping crew in the world, which makes it extremely difficult to detect undesirable elements since these states are also home to separatist movements and Islamic fundamentalism.

There are other concerns linked to the reliance on foreign-flagged vessels. During hostilities, these vessels may not be available for transporting energy imports due to the risks involved and the attendant manifold increase in insurance premiums. Besides, even if they are made available, it is not possible under international law for Indian warships to escort tankers/carriers flying foreign flags. The same is also applicable to a peacetime scenario. In case of heightened insecurity to international shipping in any area due to piracy or terrorism, this narrows down the policy options of the Indian Government in terms of employing naval escorts for these vessels.

India's domestic energy infrastructure in its maritime zones is also at risk, since the offshore platforms/rigs and their supporting vessels and infrastructure are vulnerable to collisions and sabotage by terrorists and other elements that would like to hurt India's vital interests.

As per international law (UNCLOS-3), a state can establish a security/safety zone up to a maximum of 500 metres all around offshore installations and artificial islands in its EEZ.[15] Within this zone, entry may be denied to all except authorized vessels. However, UNCLOS-3 was signed in 1982 after three decades of negotiations and hence some of its provisions may not address the requirements of the current time. Such is the case for this provision. Five hundred metres of 'cushion' is grossly inadequate to ensure the security of offshore platforms today, considering that readily available modern commercial off-the-shelf (COTS) technology has significantly enhanced the capabilities of non-state actors, for example in terms of the high speeds of attack boats, effective communication for better coordination, the increasing lethality of weapons and explosives, and even the easy availability of scuba gear and underwater scooters to carry out sub-surface attacks.

The way ahead

Explore 'swapping' options

India will soon be importing energy resources from Eastern Asia, and thus transiting the insecure straits of Southeast Asia seems to be unavoidable for the India-bound tankers. There could be ways however to avoid it, at least to some extent. In the search for a safe and unimpeded passage for shipping that is transporting fossil-fuel resources to India, one good option for India is to swap its energy imports that are sourced from West Asia (and even Africa) with those of the states in Eastern Asia. Although many of the LNG imports of energy-hungry economies of Eastern Asia are sourced from within the region (particularly from Indonesia, Malaysia and Australia), much of their imported crude oil is sourced from West Asia and Africa. It may thus be worthwhile to explore the options and feasibility (such as the compatibility of the type of oil with existing refineries) for entering into arrangements to exchange Indian oil imports with those bound for China, Japan, South Korea or Taiwan. This would not only obviate the necessity for tankers to transit insecure stretches of sea lines, but would also reduce transportation costs significantly. Of course, this would be a part solution and not a comprehensive one.

Secure energy assets overseas

'Safeguarding Indian energy assets outside territorial India' is stated in the Indian Maritime Doctrine as one of the 'scenarios in which a navy like India's can find itself – in conflict or in near conflict' (See MoD 2004).

It may be difficult to conceive a scenario wherein India would decide to use its military forces to conduct expeditionary operations against a state to secure its energy assets and stake overseas there. However, Indian forces may need to be prepared to provide security assistance to the countries that do not have the capacity to stabilize the security situation and that may request India's involvement. In the case of an eventuality that might impinge on India's energy security, the role of Indian maritime forces would be crucial, not only to transport men and material overseas to stabilize the situation ashore, but also to provide security to the offshore energy infrastructure and transportation.

Showcasing the military capabilities of Indian forces in a benign manner and providing security reassurance to the governments in these countries thus becomes an important aim of the periodical presence of Indian warships in foreign ports on goodwill missions, and naval power thus discharges its role as an instrument of foreign policy. Their mere presence close to foreign shores leads to the spread of the nation's influence and political will to secure its vital interests, and this in itself provides a measure of security.

Secure sea lines

'To safeguard India's mercantile marine and sea-borne trade in our SLOCs, both during peace and war' is stated as a naval mission in the Indian Maritime Doctrine (MoD 20040. Particularly since the early 1990s, India's trade and energy interests have been expanding far and wide. While this led to the imperative for the Navy to acquire 'long-legged' blue-water capabilities, there were severe constraints. The UNCLOS-3 also came into force about the same time (1994), which bestowed on India the sovereign rights over an extensive maritime zone. At that time the Indian Coast Guard was still a relatively young service, not sufficiently capable to tackle the low-intensity threats in the EEZ. Therefore, within the naval force levels, 60 per cent of vessels needed to be dedicated for 'brown-water' or constabulary tasks. Recent reports, however, indicate that this is now being reversed, with the ratio of 'blue water' to 'brown water' ships of 40 to 60 being turned around to 60 to 40.[16]

This, however, does not in any way indicate that the significance of the Indian Navy's constabulary role is diminished; it has actually gained greater relevance, but this role will now have to be discharged at greater distance from the shores. Due to the increasing significance of commercial shipping for national development, securing sea lines from piracy and maritime terrorism has become a prominent peace-time function for coast guards and navies worldwide. While merchant vessels could be targeted anywhere, they are most vulnerable at sea line stretches where mercantile traffic density is high. The key choke points of Hormuz Strait, Bab-el-Mandeb and Malacca Straits are thus particularly vulnerable. The surveillance and presence missions of the Indian Coast Guard and naval units do provide a measure of security to the sea lines in the northern Indian Ocean. This will be supplemented further south by the hi-tech radar stations being set up by the Navy at key locations like the Maldives[17] and Madagascar[18] to monitor mercantile traffic. Beyond the choke-points too, the Navy maintains an occasional presence as a spin-off from its foreign port visits and joint exercises with other navies. However, such measures cannot provide a comprehensive security cover. Securing the vastly extended sea lines would necessitate a prohibitively large number of assets. Major war vessels and frontline surveillance aircraft cannot be used, since operating costs are very high and it would lead to material degradation of sophisticated assets, adversely affecting their operational preparedness to perform their primary functions.

Inter-state security cooperation thus becomes an imperative. The successful recovery of the hijacked Japanese vessel *Alondra Rainbow* in 1999 by the Indian Navy and Coast Guard underscores India's commitment towards such joint efforts. To further such cooperative security in the Indo-Pacific region, in June 2006 India ratified the Regional Cooperative Agreement for Anti-Piracy (ReCAAP).[19] Initiated by Japan, this arrangement came into force on 4 September 2006. It aims to facilitate intelligence exchange through its Information Sharing Centre (ISC) being set up in Singapore. Since 2004, the Indian Coast Guard has been participating in the Asia Maritime Security Initiative (AMARSECTIVE), also initiated by Japan.[20] It seeks to garner the efforts of regional coast guards to counter non-traditional maritime threats.

It is important to note that the sea lines transiting the Indian Ocean are of immense significance to many extra-regional powers for their energy needs. Of the 200 vessels crossing the Indian Ocean every day, 40 are tankers. Sourced primarily from the Persian Gulf and Africa, the oil and natural gas laden on these tankers are vital to satiate the growing energy appetites of West and East Asian countries. The compulsions of energy security have led even Japan to extend its naval reach across the Indian Ocean, notwithstanding its 'pacifist' Constitution. Similar compulsions are leading to China's naval forays into these waters. This implies the interests of these extra-regional states converge with those of India in terms of security of energy sea lines. It also explains why, in recent years, the maritime forces of countries like the United States, the United Kingdom, France, Japan, South Korea and China have intensified their interactions with the Indian Navy and Coast Guard. Strategic analysts say that this is due to the increased salience of the Indian Ocean, but many among them also attribute the increased salience of the Ocean to the so-called 'rise' of India. Donald Berlin of the Asia-Pacific Centre for Security Studies (Honolulu) is one of them. He adds that 'New Delhi's (increasing) interest in affairs of this Ocean ... will be accompanied by a growing interest by others – especially major states – in these waters, either to check India or to ally with it' (Berlin 2002). On its part, India would need to be on the lookout for avenues of cooperation with these major navies, and constantly re-appraise these avenues, given the fluidity in regional geopolitics.

The United States is keen on an Indian involvement in the security of the Malacca Straits, as was indicated by the US Pacific Fleet Commander during his May 2006 visit to India.[21] The Chinese Ambassador to India also stated in October 2005 that 'as far as India is concerned, we don't have any problem (with its naval ships patrolling the Straits)' (from Laskar 2005[22]). There are indicators that Japan and Singapore are also amenable to India's security role in the Straits. However, Indonesia and Malaysia are reluctant to see the presence of any extra-littoral (including Indian) maritime forces in the Straits since the two states perceive it as a violation of their sovereignty. Nonetheless, India's 'will' to assist in Straits security was expressed by the Indian Defence Minister at the Fifth Shangri La Dialogue in June 2006. He said, '(S)ubject to the desire of the littoral states, as a major state-user, India would be willing to assist the (Malacca strait security) project in whatever capacity is deemed suitable.'[23] The southern part of India's Nicobar island chain is geographically well placed to enable this, since it is located only 160 kilometres from the northern tip of Indonesia's Sumatra Island, viz. virtually overlooking the northern entrance to the Straits. In any case, the ongoing coordinated patrol between the Indian and Indonesian navies along the common maritime boundary here (details in the following sub-section) provides a measure of security at the entrance to the Straits, since the patrol area is coincident with the six-degree channel, which must be transited by nearly all vessels bound for the Malacca Straits.

Any Indian role must, however, be perceived as benign and should take into account the sensitivities of Indonesia and Malaysia over sovereignty issues. Various options in that direction need to be explored. For example, India could provide assistance in capacity building through the training of local maritime forces. During the August 2005 launch of coordinated air-patrols over the Malacca Straits, called 'Eyes in the Sky', the Malaysian Defence Minister said, 'The governments outside the region could contribute aircraft and other equipment for the air patrols, (provided that these aircraft) remain in control and command of the littoral states' (from Yoong 2005). India could thus offer assistance for aerial surveillance if its air assets based in the Andaman and Nicobar Islands could be augmented and the aviation infrastructure in Campbell Bay (southern Nicobar near Indira Point) upgraded. Another way would be to have Indian Coast Guard vessels patrolling the Straits with law-enforcement officials of the littoral states on board.

Considering the importance of the Persian Gulf sea lines for India's energy imports and the fact that this region is only 600 nm from Indian shores, India has an important role to play in its maritime security. The Indian Navy does not participate in the US-led coalition Maritime Security Operations (MSO) in the waters off the Arabian Peninsula. However, as a participant in the IISS-sponsored Gulf Dialogue (which includes maritime security issues) [24] and through bilateral engagement with the sub-regional states, India needs to be continuously involved.

During a war involving India, its strategic imports would invariably be carried on board Indian-flagged carriers and provided with naval escorts. This means that all-out efforts are needed to increase the numbers and capacity of Indian-flagged vessels carrying strategic energy resources. Even during wars not involving India, these measures may be taken for critical cargo in the key sea line stretches affected by a conflict, particularly in geographic choke-points like the Malacca Straits. In order to meet adverse eventualities during wartime and to deter potential adversaries, India needs sea-control capabilities, which would necessarily include aircraft carriers, to defend shipping, carrying its energy imports, against missile and torpedo threats posed by the adversary's aircraft and submarines.

Finally, there is a pertinent question for Indian policy makers. *In a hypothetical scenario wherein navigation across the Malacca Straits or Strait of Hormuz is impeded due to terrorist action or a military conflict, would the Indian maritime forces be tasked to secure India's vital energy supplies, either alone or within a coalition of states?* It may be necessary to decide upon the responses to such questions in advance if India wishes not to be caught unawares and is thus able to take informed decisions to further its national interests.

Regulate maritime zones

Towards achieving 'good order' in India's maritime zones in terms of combating piracy and other maritime crimes, the foremost imperative is enhanced *maritime domain awareness*. This needs intensified surveillance at sea, and data linking of the various government agencies involved, from intelligence agencies and local police to maritime forces. This would enable security forces to compile a common surface picture and thus take informed and timely actions to mitigate the threats. Such domain awareness would need to be backed by patrol forces actually present in the vicinity or readily available at call to interdict the miscreants and act as a deterrence factor.

In the past few years, concerted efforts have been made in this direction, including the augmentation of aerial surveillance off the Andaman and Nicobar Islands and the southern tip of peninsular India through the induction of Unmanned Aerial Vehicles (UAV), as well as the installation of fixed radar chains and real-time digital data-links connecting the Maritime Operation Centers (MOC) with the units at sea. Although the Indian Navy's force level mix is being reversed (as mentioned earlier) with 40 per cent of its assets optimized for constabulary tasks, this policy is conjoined with major plans for the augmentation of the Coast Guard assets to make it capable of securing India's expanding maritime zones.[25] The responsibility for the security of offshore platforms, which is essentially a coast guard function, continues to rest with the Navy. For this purpose, the Indian Navy has designated a two-star Admiral as the Flag-officer Offshore Defence Advisory Group (FODAG), who coordinates the activities of various agencies towards ensuring the security of India's offshore rigs. This responsibility would eventually be transferred to the Coast Guard, when it emerges from its current constraints. This would provide the Navy greater leeway to focus on its primary blue-water role. However, irrespective of which is the lead agency for the security of India's offshore assets, adequate resources would need to be allocated for their security, particularly in terms of underwater surveillance systems.

As in the case of the security of sea lines mentioned earlier, inter-state security cooperation is essential to regulate the maritime zones. Information exchange with maritime neighbours and coordination of naval patrols along common maritime boundaries has long been considered necessary. Hence, in 2002, the Indian Navy instituted an arrangement with its Indonesian counterpart codenamed IND-INDO CORPAT, which involved constant communication exchanges among the operation centres and units at sea (ships and aircraft) of the two sides. It provided for miscreants apprehended across boundary lines to be handed over to their respective security forces. In 2005, the Indian Navy signed a similar Memorandum of Understanding (MoU) with the Thai Navy (see Suryanarayana 2005), and the patrols commenced in March 2006. Owing to these coordinated patrols, there has been a reported drop in incidents of piracy in the Bay of Bengal.[26] It may be worthwhile to enter into such arrangements with Myanmar and Bangladesh too. While this would regulate India's maritime zone in the northern Bay of Bengal, it would also provide a measure of security to Myanmar's offshore installations, from where India's natural gas may be sourced in the future.

Enhance capacity/security of ports and shipping

The Indian Government has initiated some measures to cater for the increasing inflow of tankers laden with energy resources. The Sagar Mala (Ocean Necklace) project envisages the development of many ports dotted along the entire Indian peninsula, including facilities for discharging energy resources. However, given the increasing volumes of sea-borne trade involving Indian ports, the increase of mercantile traffic may soon outpace these measures and thus advanced technological means such as the Vessel Traffic Management System (VTMS) will need to be installed in the approaches to ports and areas allocated for tankers.

While using modern equipment to increase the speed of oil and LNG cargo discharge has its economic dividends, it is also directly related to the security of the vessels, since they are more vulnerable during the turn-around period. The security of sea-ports is handled by the port operators, whereas the seaward security is the responsibility of the respective coastal states. Considering that port operations are being privatized, it is very important for seaward defence to be intensified in various ways, including through the installation of underwater surveillance equipment. The institution of the Marine police under the jurisdiction of respective states is another positive step to enhance security along the coast and port areas.

The 'Tonnage Tax' regime has been a positive step to reduce the tax liability of ship-owners and encourage them to register their ships in India, but has not been sufficient. More incentive may need to be proved to the ship-owners.

Conclusion

In recent times, states have been increasingly turning towards the sea to satisfy their vital economic interests. In terms of energy security, this means the increasing importance of oil and natural gas sourced from its extensive maritime zones and from overseas imports. This is even more applicable to countries like India that tread on a path of accelerated development.

As a corollary, those elements, both states and non-states that seek to harm these vital interests are likely to operate at sea, and the most attractive target for them would logically be the energy lifelines. While the likelihood of military conflicts has largely receded on a global scale in the post-Cold War era, the potential persists in the Indo-Pacific region. Owing to the uncertainties of the future and the need for a favourable balance of power, states are being driven to build sea-denial capabilities. The threat from non-state actors is, however, more pronounced in contemporary times. While hitherto mild 'Asian piracy' has lately become more violent and 'organized', global terrorism has emerged in the post-9/11 era and its links with the 'jihadis' operating against India are now well established. The terrorists may not yet possess adequate nautical expertise to conduct *widespread* attacks at sea, but achieving this eventually is not outside the bounds of reasonable possibility, especially considering the proliferation of a multitude of maritime crimes, piracy in particular. Besides, the terrorists have strong motivations to overcome the hurdles, and as 9/11 demonstrated, their tendency for patient and deliberate planning is well known and their ingenuity is beyond doubt.

This translates into the imperative for India to reduce its vulnerabilities with respect to its growing need for fossil-fuel resources. National efforts in this direction would necessarily include beefing up the country's intrinsic defensive mechanism and attaining self-sufficiency of vessels transporting critical energy resources from abroad as well as those sourced from India's maritime zones. Also necessary is the framing of deliberate policy, in conformity with the regional geopolitical and strategic scenario, to enhance the capabilities of maritime forces, and to forge security bonds with countries that could facilitate India's efforts towards ensuring its energy security in the coming years.

Notes

1 This chapter is a revised version of the paper presented at the TERI-KAF Conference on 'India's Energy Security: foreign, trade and security policy contexts', 29–30 September, 2006.

2 Article 55–60 of the United Nations Convention on the Law of the Sea (UNCLOS-3).
3 'Oil field auctions get 165 bids', *Times of India*, New Delhi, 16 September, 2006, p.17.
4 'India for buying LPG on term contract from Algeria', *The Hindu*, 23 April, 2007. Online: http://www.thehindubusinessline.com/2007/04/23/stories/2007042303810300.htm
5 Based on information provided by Captain Mohedin Shahul Hameed, Senior Executive, Malaysia LNG Sdn Bhd, Petronas.
6 'Sakhalin oil for India by Oct', *Times of India*, New Delhi, 18 August, 2006, p. 12.
7 'India plans to liquefy Sakhalin gas', *Reuters*, *The Moscow Times*, 24 July 2006 at http://www.themoscowtimes.com/stories/2006/07/24/049.html.
8 Tenth Five Year Plan estimates (2002–07), Planning Commission (Government of India).
9 'Piracy on the rise off Somalia', *BBC News*, 8 November 2005 at http://news.bbc.co.uk/2/hi/africa/4415196.stm.
10 'Station airs alleged Osama tape', *CBS News*, 6 October 2002 at http://www.cbsnews.com/stories/2002/09/13/attack/main522015.shtml.
11 'Suicide boat bombers attack key Iraqi oil terminal', *Jihad Watch*, 24 April 2004 at http://www.jihadwatch.org/archives/001704.php.
12 Al Qaeda is also reported to have sent a team to LTTE to sharpen its skills after the not-too-successful USS *Cole* bombing.
13 'Sea piracy in Indian waters declining', India eNews report, 3 August 2006 at http://indiaenews.com/2006-08/17235-piracy-indian-waters-declining.htm.
14 'Piracy dips but Indonesia remains violent black spot', *The New Zealand Herald*, 9 February 2005, at http://www.nzherald.co.nz/index.cfm?c_id=3&ObjectID=10010024.
15 UNCLOS-3, Article 60 (Artificial islands, installations and structures in the exclusive economic zone)
16 'Indian Navy reveals ambitious expansion, indigenisation programme' *India Defence* news report, 5 September 2006 at http://www.india-defence.com/reports/2460.
17 'India wants Maldives as intelligence base', *Minivan News*, 27 April 2006 at http://www.minivannews.com/news/news.php?id=2033.
18 'Indian Navy to lease station in Madagascar', *India Defence* online, New Delhi, 15 February 2006 at http://www.india-defence.com/reports/1357.
19 'ReCAAP would be operative from Sept 4: (Singapore) Foreign Ministry', *Web India 123 News*, 21 June 2006 at http://news.webindia123.com/news/Articles/Asia/20060621/369322.html.
20 'Coastguards adopt Amarsective 2004', *The Star*, 28 June 2004 at http://202.186.86.35/maritime/story.asp?file=/2004/6/28/maritime/8288099&sec=maritime.
21 'US sees greater Indian role in Malacca Straits', *India eNews*, 23 May 2006 at http://indiaenews.com/2006-05/8872-sees-greater-indian-role-malacca-straits.htm.
22 See also 'Maintaining security in Malacca strait by Michael Richardson' at <http://www.mima.gov.my/mima/htmls/mimarc/news/newsflash_files/news-cut/jan06.htm>
23 'India ready to help protect Malacca Strait', *Daily Times* online, 4 June 2006 at http://www.dailytimes.com.pk/default.asp?page=2006%5C06%5C04%5Cstory_4-6-2006_pg4_15
24 'The Gulf dialogue', *IISS News*, Winter 2005, pp.3–4 at http://www.iiss.org/showdocument.php?docID=712.
25 Dharmendra Tiwari, 'Coast guard gets more teeth', DNS *India* news report, September 13, 2006 at http://www.dnaindia.com/report.asp?NewsID=1052864
26 'Sea piracy in Indian waters declining', *India eNews*, 3 August 2006 at http://indiaenews.com/2006-08/17235-piracy-indian-waters-declining.htm.

References

Aziz, I., 2006. 'Gulf oil and India's energy needs.' *India Strategic*, 1 February, p. 48.

Bhaskar, U., 2006. 'China's GDP and Asian strategic matrix.' *The Economic Times*, 3 January.

Bagchi, I., 2006. 'India's fight against terror symbolic?' *Times of India*, New Delhi, 11 September, p. 12.

Berlin, D., 2002. 'Indian Ocean redux.' *Journal of the Indian Ocean Studies*, 10(1), p. 28.

BP, 2006. *BP statistical review of world energy*. British Petroleum.

Choudhury, R. R., 1998. 'Energy security policy for India.' IDSA *Strategic Analysis*, 21 February, p. 1675.

Cruz, Arlyn de la, 2004. 'We bombed ferry, claims Abu Sayyaf.' *The Nation*, 1 March.

EIA, 2007., *Country analysis briefs: Persian Gulf region.* Energy Information Administration, US Department of Energy, Washington D.C. Viewed 15 May, 2008 <http://www.eia.doe.gov/emeu/cabs/pgulf.html>.

Elegant, S., 2004. 'Dire Straits', *Time Asia*, 29 November, 2004. Viewed 15 May, 2008 <http://www.time.com/time/asia/magazine/article/0,13673,501041206-832306-1,00.html>.

Ghosh, P. K., 2006. 'The maritime dimensions of India's energy security calculus.' *Maritime Affairs*, 2(1), p. 32.

Gunaratna, R., 1998. 'Trends in maritime terrorism – the Sri Lankan case.' *Lanka Outlook*, Autumn 1998, p. 4.

IEA, 2005. *Findings of recent IEA work 2005*. OECD/IEA, Paris, France, p. 73. Viewed 1 June, 2008 <http://www.iea.org/textbase/nppdf/free/2005/findings.pdf>.

Laskar, R. H., 2005. 'No problem if India patrols regional waters: Chinese envoy.'*New Kerala* online, 28 October.

Luft, G. and Korin, A., 2004. 'Terrorism goes to sea.' *Foreign Affairs*, Institute for Analysis of Global Security (IAGS), 83(6). Viewed 1 June, 2008 <http://www.iags.org/fa2004.html>

Mitra, A., 2004. 'Indian shipping on a new wave.' *The Hindu Business Line*, 6 December. Viewed 1 June, 2008 <http://www.blonnet.com/2004/12/06/stories/2004120600440600.htm>.

MoD, 2004. *Indian Maritime Doctrine*, Integrated Headquarters, Ministry of Defence (Navy), 25 April, p. 59; 101

Suryanarayana, P. S., 2005. 'India signs maritime accord with Thailand.' *The Hindu*, 21 May. Viewed 1 June, 2008 <http://www.hindu.com/2005/05/21/stories/2005052104421200.htm>.

Tønnesson, S. and olås, Å., 2006. 'Energy security in Asia: China, India, oil and peace.' *PRIO Report*, International Peace Research Institute (Oslo), p. 40.

UNCTAD, 2000. Annex III (a) in *Review of Maritime Transport 2000*, UNCTAD Report: UNCTAD/RMT (2000)/1, United Nations Conference on Trade and Development, Geneva.

UNCTAD, 2005. Annex III (a) in *Review of Maritime Transport 2005*, UNCTAD Report: UNCTAD/RMT (2005)/1, United Nations Conference on Trade and Development, Geneva.

Yoong, S., 2005. 'Air patrols set for Straits of Malacca.' *The Standard*, 3 August. Viewed 1 June, 2008 <http://www.thestandard.com.hk/stdn/std/World/GH03Wd08.html>.

Zora, P. and Woreck, D., 2005. 'India joins the scramble for oil.' ICFI World Socialist Website (WSWS), 12 April. Viewed 1 June, 2008 <http://www.wsws.org/articles/2005/apr2005/ind1-a12.shtml>.

9 Energy security and Indian foreign policy[1]

C. Raja Mohan

Introduction

From global warming to power plays in the Persian Gulf; nuclear non-proliferation to sub-regional cooperation in South Asia; and from the politics of pipelines in inner Asia to defending sea lanes in the Indian Ocean, energy issues have emerged at the very top of India's foreign policy agenda in recent years. This has resulted in a range of new institutional mechanisms within various ministries and various inter-agency task forces. Barely a decade ago, 'energy security' as a notion hardly figured in India's foreign policy discourse. So long as the Indian economy remained a closed one and growth rates remained low, commercial diplomacy was largely alien to India's foreign policy. To be sure, India was interested in economic aid from the major Western donors as well as global multilateral institutions. The salience of this economic engagement was limited to bridging what used to be called the 'hard currency gap' and was largely the province of India's finance ministry rather than of its external affairs ministry. The foreign office was indeed quite active in the debate on the New International Economic Order (NIEO) that was once emblematic of India's foreign policy that was obsessed with leading the Non-Aligned Movement and burdened by the ideology of 'third worldism'. The emphasis was on changing terms of trade and reconfiguring norms of international economic cooperation. Although India's championship of NIEO got much scholarly attention, it had little effect either on its own economy or on the world.

It must be recalled, however, that energy-related issues did gain some importance in the diplomatic work of the nation, especially after the oil shock of 1973–74. Although India strongly supported the 'resource nationalism' of the developing world and saw the rise of OPEC as a big blow for Third World ideology, it generated little solidarity among the non-aligned nations. India's emphasis was on the narrow dimension of negotiating oil purchase contracts at the lowest possible prices from friendly oil producers. New Delhi had often leveraged its special relationships, for example with the Soviet Union and Iraq, to ensure a reliable supply of petroleum at reasonable prices. The Indian Navy was an important exception to the general lack of interest in the notion of energy security. Alone among the many Indian agencies, the Navy was naturally focused on protecting the sea lanes of commerce, and began to highlight the growing importance of ensuring the flow of hydrocarbons to India. As the weakest of the three services in a country that was preoccupied with defending its land borders, the Navy had little influence on India's security discourse.

It was only with the launch of economic reforms at the turn of the 1990s, that India began to concern itself with commercial diplomacy. As it looked out to the world in search of external investments and markets, selling India as a 'big emerging market' became one of the priorities for the Foreign Office. The Indian missions that once had seen their primary purpose as political reporting were now showing ever greater interest in economic diplomacy. (For a discussion of the emergence of economic diplomacy in India's foreign policy, see Dixit 2003.) The acceleration of India's economic growth rates since the mid 1990s saw the slow but certain emergence of a new debate on energy-related issues in India's national security discourse. The annual reports of the Defence Ministry were the first to highlight the importance of energy security. The public discourse, too, began to respond to the new concerns (for example, Raja Mohan 1997). Eventually energy security also began to figure in the concerns of major political parties and leaders. Jaswant Singh, who later held the charge for many of India's ministries during the BJP's rule (1998–2004) – finance, planning commission, defence and external affairs – was among the first political leaders to highlight the importance of energy politics in India's national security calculus (Singh 1999).

Interest in energy security began to acquire a new salience from the late 1990s amidst the dramatic rise in the volumes of India's hydrocarbon imports, the recognition that self-reliance is not an option in the energy sector, and an increasing awareness of China's dynamic energy diplomacy that buttressed the 'go out' strategy. This chapter will not address the structure of India's energy consumption, the imbalance between internal resources and external dependence, and the future of energy choices for India, as those subjects are dealt with adequately elsewhere in this volume. I simply assume that India's dependence on external energy and other resources will dramatically increase in the coming decades. Rather than focus on specific details, particular regions, or specific powers, I seek to explore the larger conceptual challenges that Indian foreign policy must confront in ensuring energy security for the nation. I would go with the definition that energy security is the successful assurance of reliable supplies of energy and related technologies at reasonable prices. I would argue that the pursuit of energy security will demand a reorientation of Indian foreign policy and diplomacy on a wide front. Energy security is not merely an additional demand on Indian foreign policy; the scale and scope of India's energy security requirements require a fundamental change in the principles that seem to have guided India's foreign policy for decades. I lay out a few broad challenges for Indian foreign policy in the context of its quest for energy security.

The logic of equity oil

The search for 'equity oil' has been the single most important new element of India's economic diplomacy in recent years. Much like China, India has made investments in hydrocarbon fields around the world a major national priority. This was reflected in Prime Minister Manmohan Singh's appeal in 2005 to the oil companies to embark on a relentless search for equity oil:

> I urge our oil and gas PSUs to think big, think creatively and think boldly in this context … They have to be more fleet-footed in making use of global opportunities, both on the supply and demand side, I find China ahead of us in planning for the future in the field of energy security. We can no longer be complacent and must learn to think strategically, to think ahead and to act swiftly and decisively.
>
> (Singh 2005)

From Sakhalin to Sudan and from Vietnam to Venezuela, India has in recent years steadily expanded its ownership of oil assets around the world. This has often demanded large scale investments by the Indian public and private sector companies, the scope of which would have been unimaginable for an Indian diplomat two decades ago. But for a rapidly growing India, now flush with foreign exchange reserves, cash is hardly the problem.

Although getting access to oil production beyond India's borders is now a well-established national objective, its underlying logic has not gone unquestioned. Some economists have argued that investing in equity oil means the squandering of precious dollars and that the current high prices of oil are not in any way a reflection of the equation between supply and demand. Instead of relying on geopolitics, the argument goes that India as well as China must rely on market forces and transparent international regulation to ensure their energy security Neither China nor India are willing to bet that market forces alone, or market forces coupled with sensible deregulation of the energy sector at home, will take care of energy security. For them equity oil is now central to an energy security strategy. Historically, non-market forces have had a great role in shaping the global energy scenario. India and China point to the US strategic policies to the Gulf in general and its relationship with the House of Saud in particular, that seem rooted in American concerns about oil security

While some experts focus on market mechanisms for energy security, others underline the reality that energy security implies a lot more than the management of supply and demand. The concentration of the world's oil reserves in a few politically volatile areas such as the Persian Gulf has highlighted the importance of power politics in shaping global energy markets. There is also a difference between the existence of energy resources and their 'availability', which could be constrained by a number of factors, including political turbulence within the major oil producing nations and the conflicts among them and between them and outside powers. As analysts have suggested, the rapid rise of oil prices during 2007–08 is related to the many conflicts in the Gulf and the premium that the market chooses to impose in this context. The traditional role of great powers in protecting their sources of energy, providing security to the regimes that own them and defending the supply routes to transport energy to the consumers, will then, remain important concerns for India in the future. The idea of 'resource wars' – of conflict among nations seeking access to natural resources, especially in the context of a dramatic rise in the demand for resources in China and India – is now upon us (for a recent debate see Victor 2007 and Klare 2008; see also Klare 2005). The Indian foreign policy discourse, however, has been reluctant to address these issues openly.

Nor have the implications of developing energy assets in the sovereign territories of other nations been fully understood in India. The dependence on equity oil in countries, often some of the most fragile in the world today, brings with it a reliance on a range of new diplomatic instruments. These include the development of non-traditional (at least for India) instruments of influence on the ruling elites or even cliques or individuals in particular countries. India has never been too good at this influence-peddling abroad, which requires approaches to money and auditing not smiled upon by the governmental machinery. India's inability to clinch many oil deals in Africa and Central Asia in the face of intense competition from China comes to mind. There have been suggestions that India has lost these contracts not because China had better bids, but because of Beijing's ability to influence the political and administrative dynamic in these countries.[2]

The challenges in relation to equity oil go beyond these normative considerations. Equity oil raises India's stakes in the stability of regimes or even individuals who preside over the resources. If access to equity oil and the protection of foreign direct investment become important objectives, the big question is how far would India go in defence of 'regime stability' elsewhere? Traditionally, India has accused great powers of discarding international norms in search of national access to energy and raw materials abroad. As India relies on vital sources from beyond its borders to ensure the economic welfare of a billion plus population, might New Delhi be compelled to adopt strategies similar to those of other great powers in defence of its national interests concerning energy security? Indian foreign policy strategists will find it difficult to evade this question for long.

The economic reforms since 1991 have led to the liberation of Indian capital and encouraged Indian business to venture abroad and take a bold global view of their operations. The increasing involvement of the Indian private sector in India's foreign policy has already been noted, in its collective lobbying effort in the United States and in the forging of new relations with major powers. The private sector is also poised to play a much larger role in India's foreign policy when it comes to energy security. New Delhi has already realized the wisdom of letting the private sector take the lead in developing the hydel potential in Nepal. The Indian decision to mobilize the services of L N Mittal in pursuing the objectives of energy security in Central Asia and elsewhere is one example of the new role of the private sector. But the future role of the private sector in foreign policy might come into full view as large Indian private energy companies like Reliance begin to make their weight felt. As companies like Reliance take on a higher profile in securing energy resources across the full spectrum– from exploration to production and marketing beyond the national boundary – Indian diplomacy for the first time will be compelled to defend private capital's foreign interests. While this may be new for India, it is normal practice in all advanced capitalist countries. But the challenge for India is to develop effective mechanisms to generate a broad set of rules and ensure their transparent implementation in this new relationship between the private sector and Indian diplomacy.

From 'Third Worldism' to 'Neo-colonialism'?

India's growing dependence on the rest of the world is not limited to oil and natural gas. India and its companies are now scouting the world coal resources. India will also need to import uranium to sustain a large nuclear power programme. Beyond energy resources, India is increasingly dependent on imports of a variety of mineral resources. As the food habits of a newly prosperous India begin to evolve, it could begin to import food on a significant scale. If an assured supply of these vital resources becomes an important national security objective for India, the nature of India's political ties with the resource-rich developing nations is bound to alter. As India enhances the national effort to gain access to resources far beyond its borders, it is beginning to confront charges of 'neo-colonialism' much like China. For the moment, Beijing is a larger target of criticism than New Delhi. China has been far more aggressive than India in pursuing its economic interests in Africa and Latin America. Western liberals and donor communities have made the point, not entirely accurately, that the hunger for resources in China and India and their incipient political rivalry is akin to the scramble for Africa among rival European colonial powers in the 19th century (for a discussion see Frynas and Paulo 2007). Irrespective of the analogy, India is certainly competing with China for oil and mineral resources in Africa. New Delhi might be way behind Beijing; but it is on the same road.

International financial institutions and major aid donors for Africa have been critical of Chinese and Indian economic policies towards Africa. They insist that Beijing and New Delhi should not repeat the mistakes of the United States and the West in bank-rolling for decades such unsavoury regimes as that of Mobutu Sese Seko in Zaire. While their aid policies are completely tied to national interest, India and China insist that they are just liberating the Third World from Western dominance. After decades of seeing themselves as victims of imperialism, China and India will find the tag of neo-imperialism and neo-colonialism shocking if not distasteful. Yet New Delhi, like Beijing, must confront a new reality. The greater their economic and political capacity to influence outcomes elsewhere in the world, the stronger will be international scrutiny of their policies. Equally important is the need to acknowledge that India's approach towards the developing world – whose resources and markets the two rising powers badly need – is undergoing a profound transformation. As high growth rates propel India, it is being compelled to design foreign policies for large economies that no longer are self-sufficient. Their rhetoric might be that of the Third World, but Chinese and Indian foreign policies could increasingly look like those of the great powers, especially in the defence of the new economic interests beyond their borders.

While foreign policy traditionalists in India recoil at the charges of neo-colonialism, realists in New Delhi must come to terms with some important factors.

One is a historically proven trend: existing great powers attack rising powers for not respecting the old rules. Although there has been a lot of talk about the implications of the global balance of power amidst the rise of China and India, Africa is perhaps the first place the new reality is being manifest. As China and India transform the geopolitical balance in Africa and threaten to push Europe and America from their privileged positions in the developing world, they are bound to face reactions from the major powers and from liberal activists in the West.

Criticism of China and India is sharpest for supporting the government in Sudan, which is facing flak on the human rights' front. Beijing and New Delhi, with their huge investments in Sudan's oil fields, have no desire to sacrifice their energy interests to compel Khartoum to change its behaviour. Support from China and India has undoubtedly emboldened Sudan to defy the international system. The same is true in Burma, where both countries are competing for influence and access to energy and other natural resources (for a broader discussion of Chinese and Indian energy policies and their specific rivalry in Burma, see Tonnenson and Kolas 2006).

To be sure, these are not the only instances where major powers have 'elevated' interests above a presumed 'principle'. India can always point to the contradictions of US policy towards the military rule in Pakistan. But India cannot have it both ways – pretending injured innocence at accusations of neo-colonialism on the one hand and asserting that other powers do the same. As its dependence on imported oil and mineral resources expands rapidly in the coming years, India, much like China, will be under pressure to defend these interests through the time-tested means employed by great powers. These include large amounts of economic assistance, subsidizing domestic capital in capturing export markets, supporting friendly governments, and selling arms to such regimes, which might use the arms against their internal and regional adversaries. In extremis, this might even involve sending troops to preserve order and stability in order to defend 'vital' national interests.

For India the challenge is two-fold. One is to immediately review current policies towards Africa and devise a strategy that treats other developing nations as partners in the march towards shared prosperity rather than as mere sources of raw materials. The other and more difficult task is to shed the old rhetoric that pretends India is a weak Third World country and start debating the consequences of India's rising power.

Use of force beyond borders

If access to energy and access to other natural resources are emerging as vital interests to Indian security, how far should India go to defend them? Should India be prepared to use force to secure these interests? These are questions that Indian analysts are loath to discuss, not just in public, but even in private. For these questions run counter to the broad national narrative that posits the centrality of norms in the conduct of Indian foreign policy and that has been hesitant to either explain or predict New Delhi's external engagement on the basis of national interest.

The talk of use of force also clashes directly with the traditional Indian emphasis on the sovereignty of developing nations and India's deep opposition to external intervention in the internal affairs of other nations (for a preliminary discussion, see Raja Mohan 2008a). But a close examination of India's foreign policy record suggests considerable ambiguity about the principle of 'non-intervention'.

Although 'sovereignty' and 'non-intervention' became important concepts in India's foreign policy discourse on a range of global issues, its neighbourhood policy had an entirely different orientation. India has often made military and political interventions in its neighbourhood, including in East Pakistan, the Maldives and Sri Lanka. (For a discussion of India's regional policy, see Mitra 2003 and Wriggins 1992.) If India has had no qualms about using force within its own region, the question is whether India will be prepared to extend the ambit of its interventions beyond the Subcontinent in pursuit of its national interests. After all, India has an impressive record of participating in international peacekeeping operations under the auspices of the United Nations (for a review, see Bullion 1997 and Krishnasamy 2003). The dispatch of forces under the UN mandate is considered legitimate and does not draw the pejorative tag of intervention. India has been comfortable undertaking UN-mandated peacekeeping operations for decades. As the nature of UN peacekeeping operations becomes more diverse, is it possible to conceive India, either unilaterally or in coalitions, working to secure either its own energy security imperatives or offering public goods for the international system? The political will to use force for foreign policy objectives within the region, and the eagerness to contribute to international security by deploying its armed forces far away from its shores, reflect an important tradition in Indian foreign policy. India's military role in regional and international security, however, has rarely got the attention it deserves in the domestic public discourse on foreign and security policy. India's emphasis on non-intervention was more about preventing the interference of other great powers in its own internal affairs and those of its immediate neighbourhood. It was not about an unflinching commitment to an abstract principle.

India's policy on international peacekeeping operations too began to come under pressure in recent years. In the past, India had insisted that a UN mandate was essential for deploying its forces beyond the Subcontinent. As the United States recognized the military contributions India could make to peace and stability around the world, New Delhi began to ease its near theological emphasis on a UN mandate. In a demonstration of its political support to the US intervention in Afghanistan after 9/11, India agreed to escort US naval ships through the Malacca Straits during 2002–03 (Khurana 2005). In 2003, the Indian Government vigorously debated the US request to send troops to Iraq. After a careful consideration of the issues involved, New Delhi eventually backed off, fearing a domestic political reaction.[3] However, that New Delhi considered such a prospect seriously was significant in itself. As the Tsunami disaster hit the eastern Indian Ocean at the end of 2004, India quickly decided to join forces with the navies of the United States, Japan and Australia in providing relief and rehabilitation (Sakhuja 2005). In June 2005, India signed a ten-year defence framework agreement with the United

States that involved a broad range of bilateral cooperation as well as participation in multinational military operations (GOI 2005). New Delhi came under sharp criticism for agreeing to join US-led military coalitions outside the United Nations. The Indian debate on using force beyond borders remains an unfinished one, but it has broken out of the restrictive confines of the past.

The question before India is not whether to intervene in the internal affairs of other states. Great powers have always intervened in the internal affairs of other nations. If India sees itself as an emerging great power, it will necessarily have to contribute to the construction and maintenance of international order. (For a discussion, see Raja Mohan 2008b). Any international order involves the use of force. For India, the real question is about defining when, where and how to intervene.

In the coming years and decades, India will have to come up with a set of guidelines for itself that will help decide which of its interests are so vital as to demand external intervention; whether such interventions should be unilateral or multilateral; and what constitutes the legitimate use of force. India would simultaneously need armed forces that are trained and equipped to undertake effective intervention and a political framework that would limit costly foreign military adventures.

Military diplomacy

India's use of force to secure access to energy and other natural resources is likely to occur only in extreme cases, if at all. But there are a range of circumstances, now described as operations other than war, where India might have to use its military to shape its energy security interests. Protecting the sea lanes of communication in the Indian Ocean is, for example, one such important area where India and its forces are already contributing in a significant manner. Energy security involves more than access to equity oil. It demands the capability to protect the transportation of energy – through the high seas – to Indian shores. Although protecting the SLOCs has always been an objective of the Indian Navy, India's growing reliance on imported energy has made it even more important in recent years (for example, MOD 2006). Protection of SLOCs, however, cannot be achieved by India's own naval capabilities. It demands cooperation with other great powers. Unstated though it might have been, India has shed its past opposition to the presence of other naval powers in the Indian Ocean, given up its demand to convert the Indian Ocean into a Zone of Peace, and has increasingly sought to work with the navies of other great powers. In the last few years, the Indian Navy has shed its past isolationism and today actively engages others through exercises and cooperative missions. But the full logic of transforming the Indian Navy into an instrument of regional security remains to unfold.

As the demands on the Indian Navy grow, its effectiveness increasingly depends on the ability of New Delhi to develop a credible military diplomacy in the region. In the last few years, India has stepped up its military diplomacy towards its neighbours. This has involved the negotiation of such new arrangements as the Defence Cooperation Agreement with Singapore in 2003 and deepening exchanges with many countries in its neighbourhood. Offering military training and arms supplies to other nations in the Indian Ocean littoral has emerged as an important diplomatic activity for India (Raja Mohan 2007a). India is also apparently looking at the options for forward deployment of some of its military assets in the region. India's alleged 'air base' in Tajikistan has drawn widespread international attention.[4] Although some of these reports might be exaggerated, there is no denying that Indian military strategy has become increasingly outward looking. At the same time, India is concerned about a similar activity by China, which is developing a 'String of Pearls' strategy in the Indian Ocean. This has involved the building of new ports in India's neighbours and in the abutting regions. (For an Indian view, see Khurana 2008.) A cynical view would suggest that as their energy demands grow and their navies are called upon to develop a secure maritime environment, India and China might well adopt behaviour similar to that of great powers in the past. Much like Britain and France in the nineteenth century and the United States and the USSR in the twentieth, India and China must be expected to place strong emphasis on access to maritime infrastructure in the Indian Ocean region. Equally important might be the inevitable development of force-projection and expeditionary capabilities for their armed forces. Such a shift would be a fundamental one in India's post-independence military thinking. But the energy imperative is promoting changes within the Indian world view and is beginning to encourage India's armed forces to think of intervention in defence of India's growing security interests beyond its borders (Kanwal 2007).

Coping with new geopolitics

The search for energy security has begun to redefine India's relations with great powers. While energy security has become a new glue in ties with the United States, Europe, Japan and Russia, it has the dangerous potential to engender geopolitical competition with China. While all the major powers, for example, have supported the historic Indo–US civil nuclear initiative, China has had reservations. Rightly or wrongly, China has tended to see the Indo–US nuclear deal in terms of geopolitics and as a potential harbinger of an alliance between New Delhi and Washington aimed at containing Beijing (Jacob 2006). Far more explicit has been India's relentless competition with China in gaining access to equity oil around the world. Sections of the Indian establishment have sought to prevent this incipient rivalry with talk of cooperation with China by seeking a joint search for equity oil and preventing an overvaluation of oil assets. But others question the prospects for success of such a venture. Sceptics point to the larger challenges that confront India and China in the realm of energy security (Kumaraswamy 2007). These include the potentially inevitable clashes between the two Asian giants in their attempt to promote their national political influences in the oil producing regions as well as in the arena of developing energy corridors; naval rivalry in the defence of sea lanes; and in the competitive development of maritime infrastructure.

As competition among major powers intersects with regional politics in oil-producing areas, the challenge of maintaining a balance of power in the Persian Gulf and Central Asia will inevitably demand Indian foreign policy attention. This will involve a major shift in the mindset of the foreign policy decision-makers – from the pursuit of preconceived ideological positions or an innocent attempt to improve relations with all players to a relentless pursuit of the national interest that might necessitate political trade-offs between competing bilateral ties with different energy suppliers.

As India's energy stakes in the Persian Gulf and Central Asia increase, New Delhi's concerns have already begun to acquire a strategic character. It will have to pay greater attention to the sources of conflict and rivalry in these two regions and the part played by global powers. In the Gulf, India will have to cope with growing resentment of the United States and the profound unpopularity of the Bush Administration in the Middle East. India is already finding it hard to balance the imperative of improving its relations with the United States with the need to manage a reasonable relationship with Iran (Pant 2007). While many in Washington have seized upon India's unwillingness to abandon cooperation with Iran as a sign of bad faith, many in New Delhi see the improvement of ties with Iran as a test case for India's foreign policy autonomy.

In Central Asia, India is already caught up in a number of competing factors (Blank 2003). The Central Asian states view Russia as both a source of regime security and as a dominant power from which greater autonomy must be sought. India must find a balance between its own separate interests in the region and its good relationship with Moscow. Meanwhile the dramatic rise of Chinese influence in Central Asia is also a matter of concern to India. The United States meanwhile hopes to integrate Central Asia with South Asia, an objective that has been reflected in the creation of a single bureau for the two regions in the State Department. While this might suit India over the longer term, its uncertain ties with Islamabad and the continuing instability in Afghanistan prevent a genuine integration between the two regions. India is also wary of the consequences of the Sino–Russian attempt to limit the US influence in Central Asia through the Shanghai Cooperation Organisation. All these trends leave India with rather difficult political choices in the Persian Gulf and Central Asia, two important energy producing regions, in the coming years.

Rethinking the neighbourhood

Within the Subcontinent, considerations of energy security are accelerating the new trend towards regional cooperation, but many enduring political obstacles remain to be overcome. While globalization has generally encouraged the South Asian states to see the value of regional cooperation, India is yet to become a vigorous champion of the Subcontinent's economic integration. (For a comprehensive review of South Asian regionalism, see Bailes *et al.* 2007.) But signs of change are indeed evident. After decades of emphasizing bilateralism, India now believes that regionalism holds the key to its own prospects for a larger role on the world stage. Its improved relations with Pakistan have breathed new life into the moribund regional organization, the South Asian Association for Regional Cooperation (SAARC) that is now committed to regional free trade as well as an energy grid in the Subcontinent. In its new quest to integrate the periphery, India is even prepared to offer unilateral economic concessions (Raja Mohan 2007b). India's enlightened self interest in promoting regionalism is strongest in the area of energy security. Besides the traditional value of access to the hydroelectric resources of Nepal and Bhutan, India has begun to recognize the importance of cooperation with Pakistan and Bangladesh to ensure the flow of hydrocarbons from the east and west of the Subcontinent. Thanks to the post-Partition political geography, Pakistan and Bangladesh had become physical barriers to India's overland trade and communications to the west and the east. The inward orientation of their economies until recently had meant that all three nations were willing to live with closed frontiers.

The new demands of energy security, however, imply that India must find a way to open up frontiers with Pakistan and Bangladesh for the overland flow of goods as well as hydrocarbons. In early 2005, India had decided to seek negotiations with its neighbours in pursuit of three pipeline projects – the TAPI pipeline to bring gas from Turkmenistan to India via Afghanistan and Pakistan, the IPI pipeline linking Iran with India through Pakistan, and a third one to bring gas from Burma to India through Bangladesh. India's decision to move forward on these pipelines involved overruling the entrenched opposition from the Indian security establishment that was loath to let Pakistan 'control' the nation's energy supplies or allow it to prosper from the transit fees that would have accrued (Pandian 2005). Pakistan, which by instinct was opposed to expanded economic interaction with India, was keen on building the pipelines as a way of enhancing its standing in the region. Pakistan's leadership, always acutely conscious of its geopolitical location at the crossroads of the Persian Gulf, Central Asia and the Subcontinent, was sensing the new possibilities of emerging as a bridge state between the three regions. Bangladesh, with less national confidence than Pakistan, remains deeply ambivalent about opening its territory for economic interaction with India. Dhaka sought Indian concessions on a range of other issues as a precondition for letting energy flows across its territory.

Decades of uneasy relations with New Delhi, a real or perceived sense of the larger neighbour's 'hegemony' and the India factor in domestic politics had made Bangladesh and Nepal squeamish about regional energy cooperation. Although the logic of self-interest demanded trans-border energy cooperation with India, the elites of the smaller neighbours were prepared to forego such cooperation in the name of standing up to India and defending national sovereignty. Nepal was reluctant to develop its hydroelectric resources for a regional market and Bangladesh found it impossible to arrive at a decision on selling natural gas to India. While the case for regional energy cooperation is now widely accepted in all the capitals of South Asia, its realization depends on the ability of these nations to depoliticize energy cooperation, separate it from the larger bilateral disputes with India, and let the private sector take the lead. This, in turn, depends on India's leadership and initiative to overcome the accumulated distrust of its neighbours.

From autonomy to responsibility: nuclear energy and beyond

Although nuclear power contributes little to India's current overall energy consumption, it could contribute in a reasonable manner to the nation's future requirements of electric power generation. But India's ability to harness nuclear power is now constrained by the international non-proliferation regime, which, by law, prohibits international nuclear cooperation with India. While India has one of the oldest civilian nuclear programmes in the world, its current strategy based on self-reliance is unlikely to generate more than 10,000 MW of electricity. As a consequence, changing India's status in the global nuclear order has become an imperative in India's energy security calculus. Amidst growing concerns about global warming and the rising prices of oil, there is a world wide resurgence of interest in nuclear energy.[5] Since the nuclear tests of 1998, gaining access to international cooperation in atomic energy has consumed much of India's diplomatic energies as well as national political debate. (For an assessment of India's nuclear diplomacy since Pokharan II in May 1998, see Raja Mohan 2006.) The culmination of this effort was the signing of the historic civil nuclear initiative between the US President George W Bush and Prime Minister Manmohan Singh in July 2005. Under the agreement, the United States agreed to change its domestic laws on non-proliferation and persuade the international community to change the global rules on nuclear commerce with India. New Delhi, in return, agreed to separate its civilian and military nuclear programmes, place the former under international safeguards, and support the global non-proliferation efforts. While much of the world saw this as an American capitulation to New Delhi's demands, within India it was seen as a potential threat.

The tortuous domestic debate on the civil nuclear initiative with the United States over the last two years reflects the domestic political difficulties of coping with the prospect of India becoming part of the global nuclear order. India's opposition to the global nuclear order made sense, so long as it was denied the right to have a nuclear weapons programme as well as access to international cooperation in peaceful uses of atomic energy. When India seems to have finally won the right to have both under the Indo–US nuclear deal, the resistance can only be explained in terms of the inertia of the old thinking and a fear of changing course. The bitter contestation, which began as a technical debate over the specific terms of the nuclear rapprochement between India and the United States, has transformed into one about the nature of India's foreign policy and its relationship with the international system.

It is interesting to note that both the right and left of the Indian political spectrum today argue against New Delhi accepting even minimal constraints on its nuclear programme or any commitments in support of global efforts limiting the spread of nuclear weapons to additional countries. For decades, India was serious in its campaign for such absolutist goals of total nuclear disarmament and quick to offer free advice to other powers on undertaking such steps as a Comprehensive Test Ban Treaty and Fissile Material Cut-off Treaty. So long as it was not pursuing a nuclear weapons programme, it was easy for India to preach the virtues of a normative approach to global nuclear issues and revel in its own sense of moral superiority. Today, there is comprehensive political opposition in India to accepting the very same measures of restraint on the testing and production of nuclear material – either in the name of national security or of anti-imperialism. Even the simplest measure of external commitment – the acceptance of international safeguards on peaceful nuclear facilities – is now questioned with great political intensity.

The official policy has accepted, rightly, some constraints on its own nuclear programme and undertaken new commitments to prevent further proliferation of nuclear weapons around the world. This was necessary not merely as an entry price into the nuclear club, but also to underline India's position as a responsible nuclear weapons power. With its own nuclear deterrent secure, New Delhi had every reason to graduate from a habitual protester against the global nuclear order to demonstrating its will to contribute to the management of that order. While the transition from autonomy to responsibility remains politically contested in the nuclear domain, the debate has not even been joined in other areas at issue concerning energy security, such as global warming and international trade. On global warming, India will find it increasingly hard to sustain the argument that it will not accept any limits on carbon emissions and those who made the mess have the sole responsibility to clean it up. While the question of equitable sharing of the cost of coping with global warming is an inherently difficult one, an obdurate focus on per capita emissions is no substitute for India eventually demonstrating leadership. As India's energy demands grow ever larger and India's choices begin to have a global impact, New Delhi needs to transform its foreign and security policies. The transition towards a more responsible foreign policy that is at once committed to securing India's national interests as well as contributing to collective goods at the global level is bound to be a long and arduous one.

Notes

1 This chapter is a revised version of the paper presented at the TERI-KAF Conference on "India's Energy Security: foreign, trade and security policy contexts", 29–30 September, 2006.

2 AFP, 'China, India fight for African Oil', *Taipei Times* (Taipei) 15 October, 2004, reproduced at http://www.energybulletin.net/2614.html, accessed on 23 May, 2008

3 For the Government of India's final statement on the troops issue on 14 July, 2003, see, <http:mea.gov.in/pressrelease/2003/07/14pr02.htm> accessed on 30 May 2008.

4 'IAF to station MiG 29s in Tajikistan', *Times of India* (New Delhi), 20 April, 2006; see also 'India has acknowledged establishing an air base in Tajikstan', *Aviation Week and Space Technology* 26 August, 2002.

5 For a quick summary of the new plans around the world to accelerate the construction of nuclear reactors, see, World Nuclear Association, 'Plans for New Reactors Worldwide', March 2008, available at < http://www.world-nuclear.org/info/inf17.html> accessed on May 30, 2008.

References

Bailes, A. J. K., Gooneratne, J., Inayat, M., Khan, J. A. and Singh, S., 2007. *Regionalism in South Asian diplomacy*. SIPRI Policy Paper No.15, Stockholm, Stockholm International

Peace Research Institute. Viewed 5 June, 2008 <http://books.sipri.org/files/PP/SIPRIPP15.pdf>.

Blank, S., 2003. 'India's rising profile in Central Asia.' *Comparative Strategy*, 22(2), pp. 139–57.

Bullion, A., 1997. 'India and UN peacekeeping operations.' *International Peacekeeping*, 4(1), pp. 98–114.

Dixit, J. N., 2003. *India's foreign policy, 1947–2003*. New Delhi: Picus Books, pp. 300–21.

Frynas, J. G., and Paulo, M., 2007. 'A new scramble for African oil? Historical, political and business perspectives.' *African Affairs* (London), 106, pp. 229–251.

GOI, 2005. *New framework for the U.S.–India defence relationship*. Press release issued 28 June, 2005 on behalf of Government of the Republic of India and the Government of the United States of America, Washington D.C. Viewed 30 May, 2008 <http://www.indianembassy.org/press_release/2005/June/31.htm>

Jacob, J. T., 2006. 'Indo–U.S. nuclear deal: the China factor.' IPCS Special Report, No. 14 (New Delhi). Viewed 30 May, 2008 < http://www.ipcs.org/IPCS-Special-Report-14.pdf>.

Kanwal, G., 2007. 'Planning for intervention: decision time for India.' *Opinion Asia*, November 29, 2007. Viewed 30 May, 2007 <http://www.opinionasia.org/PlanningforIntervention>.

Khurana, G. S., 2005. 'Cooperation among maritime security forces: imperatives for India and Southeast Asia.' *Strategic Analysis* (New Delhi), 29(2), pp. 295–316.

Khurana, G. S., 2008. 'China's "string of pearls" in the Indian Ocean and its security implications.' *Strategic Analysis* (New Delhi), 32(1), pp. 1–39.

Klare, M. T., 2005. *Blood and oil*. New York: Holt Paperbacks.

Klare, M. T., 2008. 'Clearing the air.' *The National Interest* (Washington DC), 2 January. Viewed 23 May, 2008 <http://www.nationalinterest.org/Article.aspx?id=16526>.

Krishnasamy, K., 2003. 'The paradox of India's peacekeeping.' *Contemporary South Asia*, 12(2), pp. 263–280.

Kumaraswamy, P. R., 2007. 'India's energy cooperation with China: the slippery side.' *China Report* (New Delhi), 43, pp. 349–52.

Mitra, S., 2003. 'The reluctant hegemon: India's self-perception and the South Asian strategic environment.' *Contemporary South Asia*, 12(3), pp. 399–418.

MoD, 2006. The Indian Navy's vision document, Integrated Headquarters, Ministry of Defence (Navy), New Delhi. Viewed 30 May, 2008 <http://www.indiannavy.gov.in/vision.pdf>.

Pandian, S., 2005. 'The political economy of trans-Pakistan gas pipeline project: assessing the political and economic risks for India.' *Energy Policy*, 33(5), pp. 659–70.

Pant, H. V., 2007. 'A fine balance: India walks a tight rope between Iran and the United States.' *Orbis* (Philadelphia), 51(3), pp. 495–509.

Raja Mohan, C., 1997., 'Energy and diplomacy.' *The Hindu* (New Delhi), 25 December.

Raja Mohan, C., 2006. *Impossible allies: nuclear India, the United States and the global order*. New Delhi: India Research Press.

Raja Mohan, C., 2007a. 'East Asian security: India's rising profile.' *RSIS Commentaries* (Singapore), 81/2007, viewed 30 May, 2008<http://www.ntu.edu.sg/RSIS/publications/Perspective/RSIS0812007.pdf>.

Raja Mohan, C., 2007b. 'India's new regionalism.' *Himal* (Kathmandu), 20(3), March 2007, available at < http://www.himalmag.com/2007/march/cover2.htm>

Raja Mohan, C., 2008a. 'Playing the great game: India's interventionist future.' *India in Global Affairs* (New Delhi), 1(1).

Raja Mohan, C., 2008b. 'India's great power burdens.' *Seminar* (New Delhi), 581, pp. 62–68.

Sakhuja, V., 2005, 'Indian naval diplomacy: post tsunami.' Institute of Peace and Conflict Studies (New Delhi), 8 February, 2005. Viewed 30 May, 2008 <http://www.ipcs.org/Military_articles2.jsp?action=showView&kValue=1652&keyArticle=1019&status=article&mod=a>.

Singh, J., 1999. *Defending India*. London: St Martin's Press.

Singh, M., 2005. *PM's inaugural address at Petrotech, 2005*, 16 January 2005. Viewed 23, May 2008 < http://www.pmindia.nic.in/speech/content4print.asp?id=69>.

Tonnenson, S. and Kolas, A., 2006. *Energy security in Asia: China, India, oil and peace*. Report to the Norwegian Ministry of Foreign Affairs. Oslo: Peace Research Institute Oslo.

Victor, D., 2007. 'What resource wars?' *The National Interest* (Washington DC), 12 November, 2007. Viewed 23 May, 2008 <http://www.nationalinterest.org/Article.aspx?id=16020>.

Wriggins, W. H., 1992. 'South Asian regional politics: asymmetrical balance or one-state dominance?' In Wriggins, W. H. (ed.), *Dynamics of Regional Politics*. New York: Columbia University Press, pp. 89–152.

Part III

Energy consumption and technology choices

10 Lifestyles and energy consumption

Mitali Das Gupta

There is a clear link between the lifestyles of individuals or households, and energy consumption patterns. The concept of environmental sustainability in India has been closely linked to some of these lifestyle choices – the food consumed, the mode of cooking, the transport used, recycling practices followed. India's low per capita carbon footprint is therefore not just due to its poverty and low per capita energy consumption, but also due to these clear lifestyle choices. However, in more recent times, with the growth in urbanization and globalization, energy consumption patterns in the more affluent segments of the household sector in India (as in some other developing countries), are gradually converging towards those of developed countries. In China, a recent study by Wei *et al.* (2007) suggests that approximately 26 per cent of total energy consumption and 30 per cent of CO_2 emissions in the country every year are a consequence of residents' lifestyles, and the economic activities that support these demands. In an era of globalization, and with huge metropolitan areas growing across the developing world, the rich are opting for more energy intensive consumption, shaped by values and aspirations introduced via the mass media. The implications of more energy intensive lifestyles are serious both for energy security and for climate security and need careful attention.

> Changes in lifestyles and consumption patterns that emphasize resource conservation can contribute to developing a low-carbon economy that is both equitable and sustainable.
>
> (Report for policy-makers, Intergovernmental Panel on Climate Change (IPCC) Working Group III)

This chapter discusses the issue of lifestyles and energy consumption in India through a focus on the transport, residential and commercial sectors.

Impact of household lifestyles on sectoral energy consumption

Households or individuals consume energy through two means – *direct* and *indirect*. Direct energy use includes the consumption of oil to run cars, electricity for household appliances, natural gas for heating and so on. Indirect energy consumption occurs at points before or after consumer end use – for example in the manufacture of vehicles, or the energy expended in running shops, movie theatres and so on (Sudarshan 2008). The energy embodied in consumer goods is generally greater than energy consumed directly and is thus difficult to measure. Urban households consume more energy indirectly than directly, as in the developed countries, while rural households consume more energy directly. But whether energy consumption is direct or indirect does not necessarily indicate whether households control the amount of energy consumed. For instance in the case of lighting, the number of lights that a household puts on, or the wattage, or the time, are determined by the household, whereas electricity consumed by a refrigerator is determined by the efficiency built into the appliance (unless the refrigerator has an energy efficiency label, in which case the consumer makes a clear choice). Similarly energy required for air-conditioning a house will depend on the nature of the construction and the location of the house. In case of transportation, the amount of fuel consumed will depend not only on the driving pattern, but also on urban planning, infrastructure, alternative transportation and so on.

Transport

The Indian transport sector consumes nearly one-fourth of total commercial energy, and almost half of total petroleum products. Increasing urbanization, mechanization and industrialization has meant that there has been a tremendous surge in the need for transportation. Between 1971 and 2004, there was a nearly nine-fold increase in road length and a nearly 40-fold increase in vehicle population. Poor urban planning and a lack of investment in public transportation have resulted in more and more people, especially in urban areas, opting for personal modes of transport and subsequently a steep increase in the energy needs for transport. Consumption of petrol and diesel grew at 7.3 per cent and 5.8 per cent per annum respectively between 1980–81 and 2004–05 (Planning Commission 2006). With improving income and changing lifestyles, many middle class citizens can now afford a car. As a result, the automobile industry, particularly the small car segment, is booming. The numbers of cars, jeeps, and taxis have increased at an average annual growth rate of 8.3 per cent, while two-wheelers exhibited the highest average annual growth rate of 14 per cent during the period 1970–2004, as shown in Figure 10.1.

A more detailed analysis of the motor vehicle population reflects a preponderance of two-wheelers with a share of more than 71 per cent in the total vehicle population, followed by cars with 13 per cent and other vehicles (a heterogeneous category that includes three-wheelers, trailers, tractors and so on.) with 9.4 per cent. From Table 10.1, it is evident how vehicle density has changed over time. From a mere four per thousand people in 1980, the number of two-wheelers in India has increased to 45 in 2003, which is a more than 10-fold increase. As far as cars and other vehicles are concerned, growth is slower. However, despite this increase, the current vehicle density in India is quite low, as compared to 580 in Germany and 808 per thousand persons in the United States[1]. In the coming years the profile of motorization is expected to witness a number of changes, driven by rising incomes,

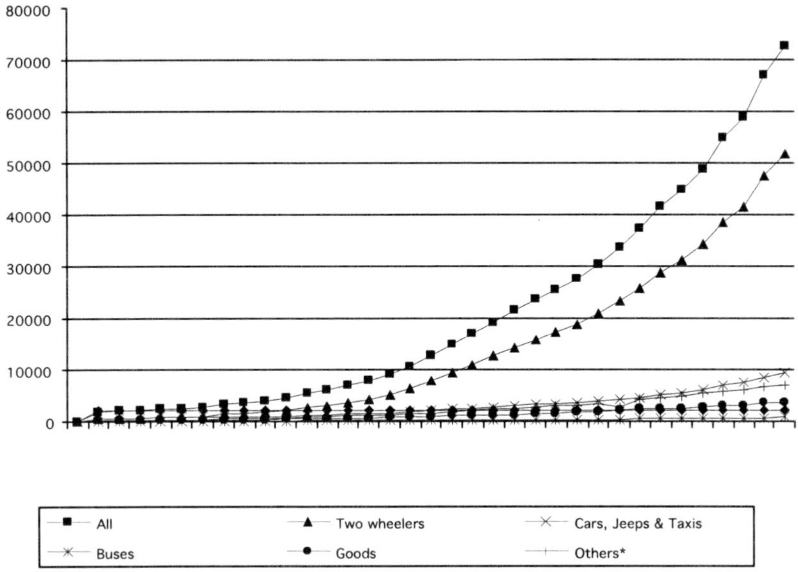

Figure 10.1 Growth of registered motor vehicles in India.

(Source: Indiastat.com. <http://www.indiastat.com/ > (accessed 2 May 2008). Data from Ministry of Shipping, Road Transport & Highways, Govt. of India.)

Note:

* Others include tractors, trailers, three-wheelers (passenger vehicles) and other miscellaneous vehicles, which are not separately classified.

desire for safety and comfort, and government regulations. Over the short term, the sensitivity of demand for vehicles to changes in GDP and in vehicle price is somewhat elastic as their purchase can often be delayed, but in the long term it has low elasticity, indicating that personal mobility is considered by many as essential to everyday living and that it has few substitutes.

On the one hand personal mobility is increasing day by day, and on the other, the cost of travel, especially for the poor, is increasing considerably. This is largely because the use of cheaper non-motorized modes of transport such as cycling and walking has become extremely risky, since these modes have to share the same right of way with motorized options. In rural areas, non-motorized modes such as pedestrians, bicycles, rickshaws, and bullock carts remain dominant, though with a growing provision of road infrastructure and changing lifestyles, the situation is changing rapidly in rural areas too. According to the International Energy Outlook 2007 (see EIA 2007), India is expected to show the largest increases in consumption of energy by the transportation sector among the non-OECD countries. The combined growth rate for transportation energy use in all the countries of the Central and South American economies is projected to be similar to that in India.

India's greenhouse gas emissions from the transport sector are relatively modest in comparison with China, Europe and the United States. Currently, the segment

Table 10.1 Number of registered motor vehicles per thousand persons in India

Year	*Two-wheelers*	*Car/jeep & taxis*	*Other vehicles*	*Total*
1980	3.7	1.6	2.2	7.6
1985	6.9	2.1	3.1	12.1
1990	15.0	3.2	4.6	22.8
1995	23.0	4.3	6.2	33.5
1996	25.2	4.6	6.9	36.6
1997	27.4	5.0	7.4	39.7
1998	30.0	5.4	7.9	43.3
1999	32.3	5.7	8.2	46.2
2000	33.6	6.1	8.5	48.1
2001	37.5	6.9	9.2	53.5
2002	40.0	7.3	9.4	56.7
2003	45.3	8.2	10.4	63.8

Source: Indiastat.com. Viewed 24 April 2008 <http://www.indiastat.com/ >.

with greatest demand is freight, unlike developed countries where emissions from private passenger transport are more significant than emissions from freight. This is expected to reverse in the next half-century or so. Emissions from passenger cars are expected to show the greatest growth in the coming decades (roughly five per cent per year) and will surpass emissions from two-wheelers, even though the latter will continue to capture the dominant market share of passenger vehicles (Figure 10.2).

The approaches necessary to reconcile the needs of the transport services and those of environmental sustainability involve two complementary routes. The first would be to make use of appropriate policies, cleaner fuels and modern technologies to encourage the flexible use of vehicle modes and thereby minimize the adverse impact on the environment, and the second would be to apply planning strategies and induce lifestyle changes to improve access to and availability of public transport and thereby reduce the need for private transport.

Residential

The residential sector in India is responsible for around 13–14 per cent of total commercial energy use (TERI 2004). Cooking is the major activity that accounts for the highest amount of direct energy needs of Indian households. Different fuels are used for cooking, and if all energy carriers are considered, this single end-use accounts for about 35–45 percent of the household energy consumption (Pachauri 2007). In the developed countries, cooking consumes less than 10 per cent of national fuel consumption. On the other hand, only 10 per cent of commercial energy that is electricity and fossil fuels is consumed in India's household sector for cooking.

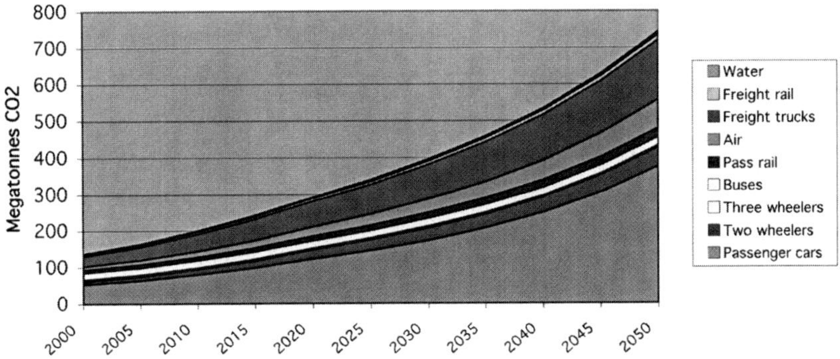

Figure 10.2 Expected growth in CO2 emissions in India from different transport modes. (Source: IEA 2004.)

Energy access is a major problem in rural India, where traditional biomass fuels such as firewood and chips, and dung cake are still the main sources of energy for household cooking. The 61st round of the National Sample Survey of India (2004–05) found that firewood and chips continued to be the most important source of energy used for cooking in rural India, with 75 per cent of rural households dependent upon it (Figure 10.3). But there is a decreasing trend in the percentage of households using firewood and chips over the period. One per cent and five per cent of rural households have switched away from this source since 1999–2000 and 1993–94 respectively. There has also been a two per cent decline in consumption in the 'others' category, which essentially includes cooking with coal and coke, biogas, charcoal and kerosene. Substantial changes have been seen in the case of LPG consumption, which increased from two per cent in 1993–94 to five per cent in 1999–00 and nine per cent in 2004–05.

In urban India, households use mainly three primary sources for cooking, namely LPG, firewood and chips, and kerosene. There has been a decline of eight percentage points in the use of firewood and chips over the years 1993–94 to 1999–00 from 30 per cent to 22 per cent as shown in Figure 10.3. This has remained so in 2004–05. LPG has gradually gained acceptance in India. Only 30 per cent of households were using LPG as a primary source of energy for cooking in 1993–94, while 44 per cent used it during 1999–00 and 57 per cent used it in 2004–2005. The consumption of kerosene has reduced drastically over 1999–2004, by 12 percentage points. It could be argued that LPG is displacing kerosene more than firewood and chips as the primary source of energy. An interesting feature is the emergence of a 'no cooking' situation in urban India, helped by the growth in hotels, restaurants and eating-houses that are available to facilitate eating out for the urban population. As far as the 'other' category is concerned, comprising dung cakes, gobar gas, electricity and coal, there has been a 5 per cent decline in their consumption by households between 1999–04. It is thus clear that in the urban areas there has been a significant shift in the fuel mix, but that the same does not

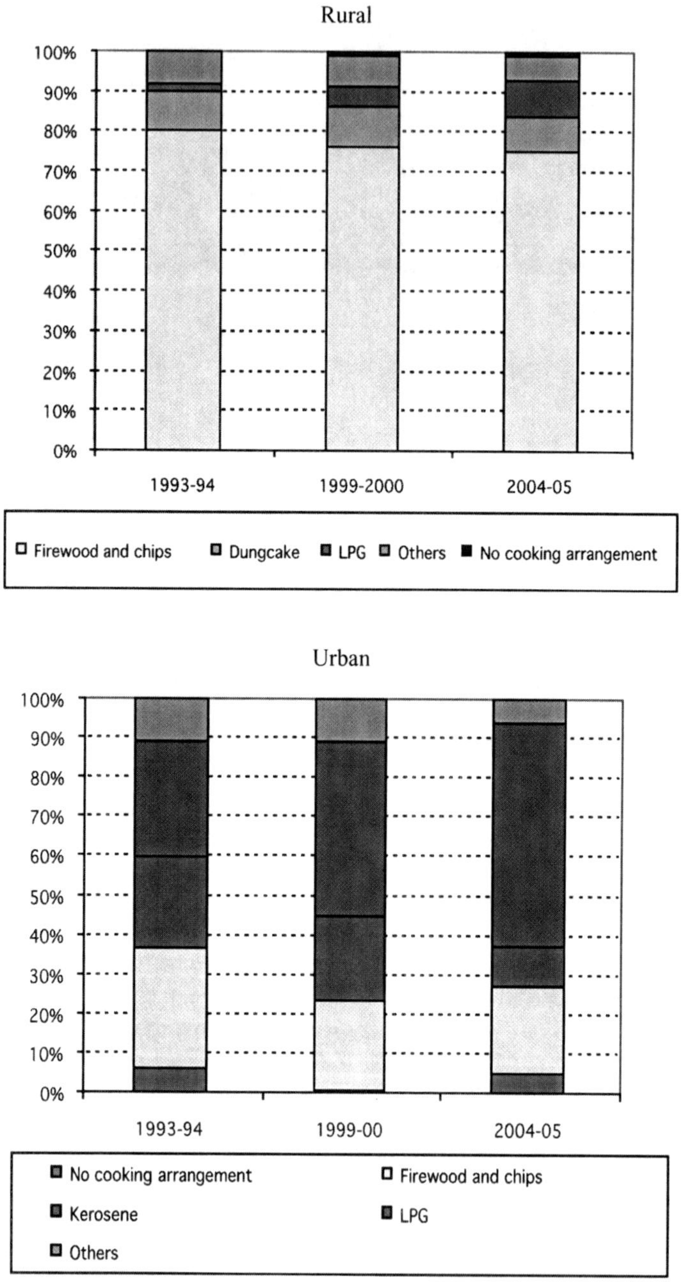

Figure 10.3 Distribution of households by primary source of energy used for cooking – rural and urban India.

(Source: NSSO 2001; 2007.)

hold for rural India. In fact the slow rates of change in rural India underline how far the country still needs to go in terms of making modern energy accessible to a majority of people.

Lighting is the second most important activity, next to cooking. For lighting purposes, households either use electricity or kerosene, accounting for about 99 per cent of usage in both rural and urban areas. However, over time electricity is increasingly displacing kerosene as a source of lighting both in rural and urban India. Between 1993–94 and 2004–05, about 17 per cent of households in rural India shifted from kerosene to electricity, as against 9 per cent in the case of urban India (Figure 10.4). In the rural areas, use of kerosene, though it has gradually diminished, constitutes a major share of household energy consumption. It is 44 per cent, as compared to only 7 per cent in urban India. On the other hand, 55 per cent of rural households use electricity for lighting as against 92 per cent in urban India.

Apart from cooking and lighting, urban households use electricity for a range of other activities like water heating, space conditioning, refrigeration, and entertainment. A Greenpeace report (Greenpeace 2007) says that the richest consumer class in India produces 4.5 times more CO_2 than the poorest class. The relatively rich consumer segment uses all sorts of modern electronic devices, ranging from DVD players, air-conditioners, and television sets to kitchen equipment, and from mobile phones to computers. Table 10.2 shows the penetration of consumer durables and white goods in India per thousand households. It shows that between 1995–96 and 2005–06, the maximum penetration has been in the case of motorcycles, followed by cars and television sets. The same trend is likely to follow in 2009–10 as well.

This suggests that the increase in electricity consumption is going to be pronounced. Letschert and McNeil (2007) project that on an average between 2000 and 2030 per capita energy consumption in India will grow at a rate of 8.2 per cent a year due to the increased penetration of white goods among households. Conversion to modern fuels will thus add to the increased energy use per capita of Indian households.

Table 10.2 Penetration of consumer durables (number of households owning goods per '000 households)

	1995–96	*2001–02*	*2005–06*	*2009–10*
Cars	16.1	30	50.2	91.4
Motorcycles	29.3	70.8	147.6	282.6
TV	72	145.6	213	314
Refrigerators	86.1	134	160.7	224.9
Other white goods	149.4	247.1	319.1	451.7

Source: NCAER (2005).

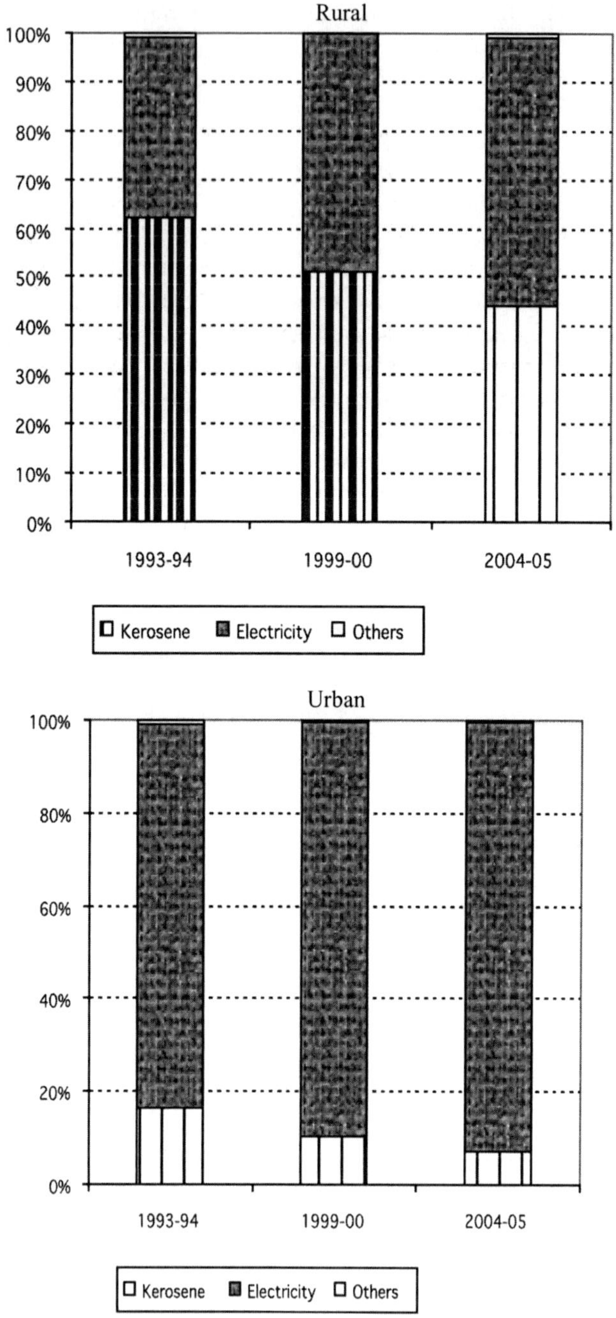

Figure 10.4 Distribution of households by primary source of energy used for lighting – rural and urban India.

(Source: NSSO 2001; 2007.)

Commercial

Though the commercial sector comprises various institutional and industrial establishments such as banks, hotels, restaurants, shopping complexes, offices and public departments supplying basic utilities, it is towards real estate (both residential and commercial) that maximum investment is directed. With the Indian government substantially easing norms for foreign direct investment in the construction and real estate development sector, the chances of the sector attracting large scale foreign investment have improved dramatically. The construction sector in India is witnessing fast growth due to several other factors as well. These are increased demand for housing, strong demographic impetus, expansion of organized retail facilities, increased demand for commercial office spaces by multinationals and IT (information technology) hubs; setting up of special economic zones (SEZs); easy availability of finance; and the increase in per capita income and standard of living. There is significant demand for quality housing in India. Many real-estate developers are developing massive township projects. With younger people wanting to own a house, with greater affordability due to reasonable interest rates on home loans and favorable tax treatment, the demand for residential housing is expected to continue to grow in the near future.

Construction materials are energy intensive. The use of appropriate materials and design can save a significant amount of energy, both in construction and use[2]. At the national level, domestic and commercial buildings account for more than 30 per cent of annual electricity consumption. TERI studies show that air-conditioning and lighting are the two most energy consuming end-uses in the building sector. With a near consistent 8 per cent rise in annual energy consumption in the residential and commercial sectors, energy consumption in buildings has seen an increase from a low 14 per cent in the 1970s to nearly 33 per cent in 2004–2005 (TERI, 2006). As per TERI estimates, assuming that the average energy consumption for a residential property is 80 kWh/m^2/annum and for a commercial building is 160 kWh/m^2/annum in the business as usual scenario, a 10 per cent annual increase in the built-up area for residential and commercial buildings will lead to a projected annual increase in energy demand of commercial and residential buildings to the tune of 5.4 billion kWh. This will be in addition to energy requirements for the manufacturing of building materials and equipment, and energy used during construction.

Figure 10.5 shows the residential sector's energy consumption in OECD and non-OECD countries. According to the EIA, for the non-OECD region as a whole, real GDP is projected to grow by more than 5 per cent per year on average from 2004 through 2030; the population is projected to grow by more than 1 per cent per year, and household energy use is projected to grow at a robust rate of 2.4 per cent per year, as higher incomes will foster increased use of energy-using appliances. As a result, households in the non-OECD nations are projected to consume about 10 per cent more energy than households in the OECD nations in 2030. China and India are expected to account for more than 40 per cent of the increase in residential energy use in non-OECD countries through 2030, as their economies continue to grow rapidly over the projection period. By 2030, the non-OECD countries will account for around 53 per cent of the global residential sector's energy consumption. This is indicative of the converging lifestyles of the relatively affluent section in the developing countries with those in the developed countries.[3]

Lessons from Japan

Since 1990, Japanese household emissions have gone up nearly 40 per cent. Some of Japan's environmental experts attribute this rise to increasing consumerism.[4] Consumer behaviour, it has been said, holds the key to Japan's ability to fulfill its commitments under the 1997 Kyoto Protocol (Kitazume 2007). However, Japan is a country where energy consciousness is quite high and a number of measures have been taken to make household lifestyles more energy efficient and sustainable.

Around 1994 in Japan, if people wanted to put solar panels in their apartments, it would cost around $60,000 (Mori 2008). Therefore in a bid to promote solar panels, the Japanese government paid half the cost of new solar installations. Sales went up and costs came down by about a third. The government phased out the subsidies gradually and ended them in 2005. Still certain local governments continue to provide subsidies to install solar panels. The price of solar panels still needs to come down by half before homeowners and builders really take the plunge to buy. Japan is aiming for 30 per cent of households to have solar panels by 2030.

The two oil crises of the 1970s had a great impact on Japan's subsequent energy policies. Japan enacted the Law concerning the Rational Use of Energy (Energy

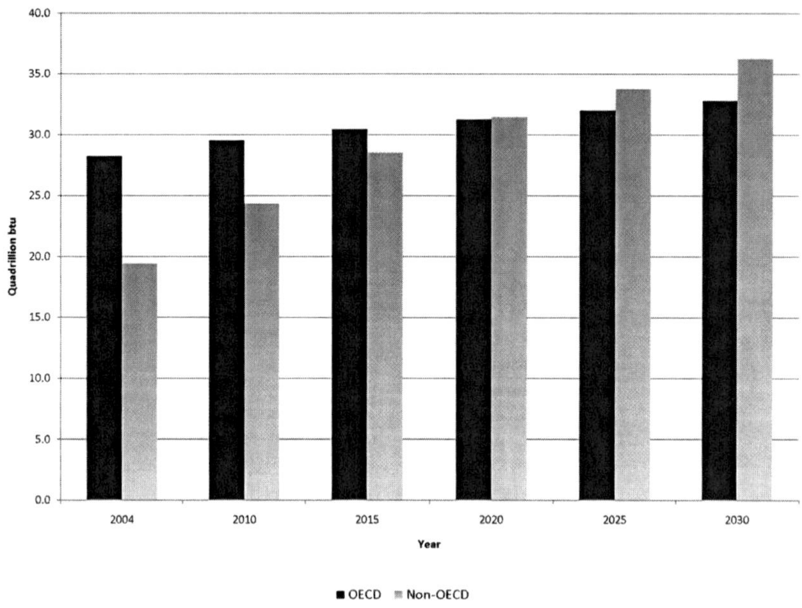

Figure 10.5 OECD and non-OECD residential sector delivered energy consumption 2004–2030 (Quadrillion Btu).

(Source: <http://www.eia.doe.gov/oiaf/ieo/excel/figure_26data.xls> accessed 11 March 2008.)

Conservation Law) in 1979, which provided a legal basis for energy conservation activities. The Top Runner Programme was developed in this context in 1999, in addition to the existing Energy Conservation Law. It applies to machinery and equipment in the residential, commercial, and transportation sectors. The programme sets fuel efficiency standards higher than the performance of the best product commercially available in the product category. Manufacturers who do not meet the standards are given advice, publicly announced, given an order, or fined (one million yen or less). Twenty-one categories of products have been covered by the Top Runner Programme since 2006 (Miki 2006). Target appliances included air conditioners, TV, refrigerators, stoves, gas heating appliances, and personal computers. As a result of this policy, significant improvements have been achieved in developing energy efficient electronic appliances. LCD and plasma TV sets and heavy vehicles have been added recently.

The energy-saving labeling system has also been introduced in Japan to inform consumers of the energy efficiency of home appliances and to promote energy efficient products. The labeling is applied to products like air conditioners, fluorescent lights, TV sets, electric refrigerators/freezers, space heaters, gas cooking appliances, gas water heaters, and computers. But these appliances have still not proliferated to any great extent among households. This is partly due to their higher prices compared to those of ordinary appliances. Switching to energy-saving electrical appliances could reduce household emissions of greenhouse gases by up to 40 per cent, according to Japan's white paper on the environment, released in June 2007. The government estimates that if all homes in Japan use fluorescent bulbs, it could cut some 2 million tons of CO_2, or 1.3 per cent of all household emissions.[5]

In April 2005, the Global Warming Prevention Headquarters, led by the Japanese government, launched a large-scale national campaign called 'Team Minus 6 per cent' in collaboration with the private sector. This aimed at providing information and raising public awareness about the issue of climate change. The campaign focuses on sharing simple tips to help prevent climate change, as some surveys show that, without knowing where to start, people are less likely to translate intent into action. The campaign aims to have individuals, businesses, and other organizations work together to achieve a 6 per cent reduction in GHG emissions. In particular, it calls on people to:

- set air conditioners at 28 degrees Celsius (temperature control)
- avoid wasting water at the tap by not letting it run unnecessarily (wise use of water)
- choose and buy energy-efficient and eco-friendly products (green purchasing)
- stop car idling (smart driving)
- say no to excessive packaging (waste reduction)
- unplug devices when they are not being used (wise use of electricity).

Apart from these efforts, the old custom of sprinkling water with a ladle on streets and gardens, called 'uchimizu' is a well known example of the use of water in Japan's daily living. People sprinkle water, especially in the summer time, in their house entrances and gardens or in front of their shops and offices to settle dust and ease the heat. Another interesting campaign called the light-down campaign started in 2003 (Mori 2008). In 2007, the campaign asked for the voluntary participation of business facilities and households in turning off the light at night over a three-day period in the month of June. The number of facilities that participated rose from 2,278 in 2003 to 63,138 in 2007. The estimated savings of energy amounted to 2.9 GWh and the estimated reduction on GHG emissions totaled 1,136 tonnes of CO_2.

There is also a nation-wide campaign in Japan to reduce indoor air conditioning by wearing clothes which make people feel much cooler in summer and much warmer in winter. The campaign called 'Cool Biz' (during summers) and 'Warm Biz' (during winters) helps office workers adapt to set room temperatures. Engaged in this energy conservation programme, many companies have reported large savings on their electricity bills. A similar initiative called 'Uchi Eco' has been launched, encouraging individuals to save energy at home by focusing on appropriate clothing, food and housing.

In the transport sector, fiscal incentives have been given to the purchase of low-emission vehicles. However, further reductions in the transport sector require more comprehensive measures. The Government has started Environmentally Sustainable Transport (EST) model projects in 27 localities throughout Japan, to pursue a number of measures like road pricing, advanced traffic management systems, car pooling, and promotion of public transport. In many cities the no-idling campaign has been promoted for public buses, by developing an automatic idling stop-start system. The white paper also expresses a strong sense of urgency for implementing energy-saving technologies, including high-performance fuel-cell batteries to help commercialize electric cars.

This discussion demonstrates that the Japanese have undertaken a number of initiatives to help move towards a more energy efficient lifestyle. Though each one of them may have only a small and sometimes invisible impact in reducing GHGs, nevertheless, in the long run such measures raise awareness among people and communities, and create significant changes in the way people use energy and other natural resources.

Policy issues in India

Transport sector

In India, policies have generally catered to the popularity of private transport among the relatively affluent section. Central and state governments offer a range of tax breaks, subsidies, and regulatory concessions that enhance the automobile industry's profitability. In addition, governments at all levels have concentrated on the expansion of roadway capacity to accommodate the increased volumes of motorized travel, especially by car and truck. In the mid 1990s, the central government greatly increased its commitment to improve the overall roadway system. In 1995, the Indian Parliament passed the National Highway Act, which established the National Highway Authority of India as well as the new Central Road Fund. These developments initiated an ambitious programme of roadway expansion and modernization. From 1997 to 2005, the extent of Indian National Highways grew from 34,298 to 65,569 km. Much of the roadway expansion in India since 1997 has been between major cities. There are also new efforts to improve highway connections between major ports and large cities. The Indian government has recently initiated an ambitious project to pave and widen much of its already existing National Highway Network (Planning Commission 2007). But even with these efforts, the Indian roadway network lags in terms of quality and carrying capacity. Since India's new and improved roadways are mostly between cities, their main impact is on intercity and interstate travel and therefore they have a smaller impact on travel within the city. While there have been considerable improvements in rail services, bus services continue to deteriorate, forcing many passengers to choose faster motorized modes such as cars and motorcycles. Although walking and cycling account for about half of all trips in the urban areas, they do not receive the funding necessary to generate quality improvements, provide support infrastructure and put in place legal rights or even traffic priority.

In India, cities are rapidly expanding into their surrounding areas. To some extent, this is the natural result of rapid population growth and the need to develop suburban areas to accommodate new housing and commercial development. However, there is a lack of coordination between such development and the provision of roads and public transport. The consequence has been rapidly rising trip distances, increasing reliance on private cars to get around, worsening traffic congestion, and mobility problems for the poor who cannot afford to live in the more accessible central city areas. Like big cities in the Asian and Latin American countries, the Indian Government too has recently introduced the BRT (bus rapid transit) system in some states to ease the traffic situation. But since the bulk of urban trips are conducted over short and medium distances, the number of motor vehicle-owning commuters who would use such a circumscribed network is likely to be limited, especially given the unmatched navigability of two-wheelers and three-wheelers in congested road conditions; ease of parking; their ability to carry passengers and luggage at low cost; and, most importantly, people's preference for personal mobility over dependence on public transport. Therefore, provision of mass transit has to go hand in hand with policies to discourage the use of private vehicles (Badami 2007). In the present scenario it is unlikely that the trend towards increased motorization and its impact on energy usage and emissions is going to reverse. Even with strong government policies, it can probably only be slowed down.

Residential sector

Relatively few policy measures have been implemented in India to cut down energy consumption (and sometimes wasteful energy consumption) in the residential sector. However, in May 2007, in order to reduce the energy consumption of the household sector, the Bureau of Energy Efficiency (BEE) agreed on two actions pertaining to a bulk procurement and distribution of compact fluorescent lamps (CFLs) and promoting and mandating the use of energy efficient pumps and other energy efficient appliances.

The scheme on CFLs seeks to replace an estimated 400 million incandescent bulbs with CFLs, which could save 6,000 MW to 10,000 MW annually. It seeks to make available high quality CFLs at the cost of incandescent bulbs. The basic premise is to create an appropriate legal and regulatory environment for energy efficient end use products, and to provide the consumer with options to make an informed choice. The plan aims to reduce overall annual energy consumption by 3,000 MW by the end of 2012. However, a major stumbling block towards promoting a sense of energy efficiency is a lack of information and awareness among households and a lack of monitoring and regulation. The problem is also acute for very poor consumers. Nearly half of the residential consumers in most states have an electricity use of less than 50 kWh/month. The monthly electricity bill of these consumers is barely Rs 50 to Rs 85. These consumers are too poor to afford CFLs at present prices – and they make up more than 50 million households. Additionally, about 100 million houses without electricity, which would be connected to the grid in the coming five years, fall in this category. Possibly India should take lessons from Japan as to how government initiatives can lead to significant energy savings in the household sector by implementing simple methods that are important in raising public awareness but do not impinge greatly on consumers' comfort levels.

Construction sector

There have been several initiatives by central and state governments to steer the construction industry to design, develop and operate sustainable buildings. The Ministry of Environment and Forests (MOEF) has mandated environmental clearance for all large construction projects above a certain size and cost. The Environmental Impact Assessment (EIA) Notification, 1994 was amended on 7 July 2004, making it mandatory for new projects relating to the construction of new townships, industrial townships, settlement colonies, commercial complexes, hotel complexes, hospitals, industrial estates, and office complexes above a certain size to obtain prior environmental clearance from the central government before starting any construction on the new project. The EIA process helps in addressing various environmental issues such as management of municipal solid wastes and industrial and hazardous wastes; air pollution arising from vehicles and traffic congestion; dependence on fossil fuels for energy demand, and conservation of water, which urban planning regulations are unfortunately not able to address comprehensively.

The Energy Conservation Act led to the formation of the Bureau of Energy Efficiency that initiated the formulation of India's Energy Conservation Building Code (ECBC). The ECBC covers minimum requirements for building envelopes; mechanical systems and equipment, including heating, ventilation and air conditioning (HVAC) systems; interior and exterior lighting systems, and the servicing of hot water systems, electrical power and motors in order to achieve energy efficiency in the different climate zones of India. The Ministry of New and Renewable Energy (MNRE) has launched GRIHA (Green rating for integrated habitat assessment), developed by TERI as a national rating system, and has plans to incentivize green buildings through this system. GRIHA has been developed for the new commercial, institutional and residential constructions that are very large energy consumers in the building sector.

The Confederation of Indian Industries (CII) and Federation of Indian Chambers of Commerce and Industry (FICCI) have played an important role in promoting energy efficiency initiatives in the country's building sector. The Indian Green Business Center (IGBC), under the Confederation of Indian industries (CII) is facilitating the LEED (Leadership in Energy and Environmental Design) rating of the United States Green Building Council (USGBC). Introduction of the LEED rating system has stimulated innovation within the building materials' supply industry.

However, even with these initiatives, there are some major implementation barriers. General apprehension of a high initial cost of constructing green buildings is the primary barrier. Second, developers do not benefit directly by incorporating energy efficient features in their construction projects. Thus, with a lack of incentives as well as lack of information for builders on how to integrate environmentally friendly features in their construction, the penetration of energy efficient technologies is very limited in the residential sector. Third, though regulations like environmental clearances and rating systems exist, there is a total absence of any monitoring mechanism. Finally, though energy audits are mandatory for many government buildings and commercial buildings above a certain threshold of connected load, there is no mandatory requirement with respect to a minimum energy performance.

Concluding remarks

This chapter examined the link between the pattern of lifestyles in India and energy consumption by looking at the current situation in the transport, residential and commercial sectors. The crucial issue that emerges is that, as development progresses and people's incomes rise, households tend to lead a lifestyle gradually approaching that of their counterparts in developed countries. Therefore, adopting sustainable lifestyles must be viewed as a necessary and complementary means of controlling energy consumption and addressing issues of energy security. The chapter highlighted some of the policy initiatives in place in Japan, which have enabled significant changes in the way people use energy at home. Finally it focused on existing policies in India and examined why they have been largely ineffective in bringing about more sustainable patterns of energy consumption.

Notes

1 National Road Transport Policy (As recommended by Thangaraj Committee), viewed 5 April 2008 <http://morth.nic.in/writereaddata/sublinkimages/ROAD_TRANSPORT_POLICY7406816237.htm>.

2 Air-conditioning and lighting are the two most energy consuming end-uses in the building sector.

3 A pilot study is under way to examine the issue of converging lifestyles and carbon footprints in urban India.

4 Details available at <http://www.npr.org/templates/story/story.php?storyId=15321013> (accessed 25 January, 2008).

5 Lights out for incandescent bulbs in energy-saving proposal - The Japan Times Online, 20 December 2007, available at <http://search.japantimes.co.jp/cgi-bin/nb20071220a1.html> (accessed 2 January, 2008).

References

Badami, M. G., 2007. 'High capacity bus system and the road beyond.' *The Hindu*, 5 December. Viewed 3 May, 2008 <http://www.thehindu.com/2007/12/05/stories/2007120553611000.htm.>

EIA, 2007. *International energy outlook*, Report No. DOE/EIA-0484 (2007), Energy Information Administration, May 2007, pp. 19–28. Viewed 1 June, 2008 <http://www.eia.doe.gov/oiaf/ieo/pdf/enduse.pdf>.

Greenpeace, 2007. *Hiding behind the poor: a report on climate injustice*. Bangalore: Greenpeace India Society.

IEA, 2004. *SMP (Sustainable Mobility Project) model documentation and reference case projections*. Paris: International Energy Agency.

Kitazume, T., 2007. 'Improving consumer lifestyle choices key to meeting CO_2 goals.' *The Japan Times*, 28 April.

Letschert, V. E. and McNeil, M. A., 2007. 'Coping with residential electricity demand in India's future – how much can efficiency achieve?' Paper No. LBNL- 63199, Environmental Energy Technologies Division, Ernest Orlando Lawrence Berkeley National Laboratory.

Miki, T., 2006. 'Energy efficiency and conservation policy in Japan.' Energy Conservation and Renewable Energy Department, Agency of Natural Resources and Energy (ANRE) and Ministry of Economy, Trade and Industry (METI), Japan.

Mori, H., 2008. 'Japan's contribution towards building a low carbon society in Asia.' *TERI/AEI Special event on lifestyles, energy security, and climate*, 8 February, New Delhi, India.

NCAER, 2005. 'The great Indian market.' Results from the NCAER's Market Information Survey of Households in association with Business Standard.

NSSO, 2001. *Energy used by Indian households 1999–2000.* NSS 55th Round, National Sample Survey Organization, Ministry of Statistics and Programme Implementation, Government of India.

NSSO, 2007. *Energy sources of Indian households for cooking and lighting 2004–05.* NSS 61st Round, National Sample Survey Organization, Ministry of Statistics and Programme Implementation, Government of India.

Pachauri, S., 2007. *An energy analysis of household consumption: changing patterns of direct and indirect use in India.* Alliance For Global Sustainability Bookseries. Berlin: Springer.

Planning Commission, 2006. *Integrated energy policy.* Report of the Expert Committee. New Delhi: Government of India.

Planning Commission, 2007. *The Working Group report on road transport for the eleventh Five-Year Plan.* New Delhi: Government of India.

Sudarshan, A., 2008. 'Lifestyles, energy consumption, and climate change: a global view of the links.' *Energy Security Insights*, 2(4), pp. 2–9.

TERI, 2004. *Teri Energy Data Directory and Yearbook 2002/03.* The Energy and Resources Institute. New Delhi: TERI Press.

TERI, 2006. *Teri Energy Data Directory and Yearbook 2004/05.* The Energy and Resources Institute. New Delhi: TERI Press.

UNEP, 2005. 'Sustainable energy consumption.' Background paper prepared for the European Conference under the Marrakech Process on Sustainable Consumption and

Production (SCP) Berlin, 13–14 December, UNEP/Wuppertal Institute Collaborating Centre on Sustainable Consumption and Production (CSCP).

United Nations Department of Economic and Social Affairs, 2007. 'Sustainable consumption and production, promoting climate-friendly household consumption patterns.' Division for Sustainable Development, Policy Integration and Analysis Branch.

Wei, Y.-M., Liu, L.-C., Fan, Y., and Wu, G., 2007. 'The impact of lifestyle on energy use and CO_2 emission: an empirical analysis of China's residents.' *Energy Policy* 35(1), pp. 247–257.

WHO, 2008. 'Reducing your carbon footprint can be good for your health: a list of mitigating actions.' *Protecting health from climate change*, World Health Day 2008. Viewed 1 May, 2008 < http://www.who.int/world-health-day/toolkit/dyk_whd2008_annex1.pdf>.

11 Technology options for India's energy security

A summary of a modelling exercise

Pradeep Kumar Dadhich

The growth of a developing economy is highly dependent on the growth of its energy consumption. Because of the possibility of inter-fuel substitution in end-use applications, the optimal supply of long-term energy requirements of a country necessitates an examination of all energy resources available both indigenously and globally. The Government of India plans to achieve a GDP growth rate of 10 per cent in the Eleventh Five Year Plan and to maintain an average growth of around eight per cent in the next 15 years (Planning Commission 2002).

Given plans for rapid economic growth, it is evident that the country's requirements for energy and supporting infrastructure would increase rapidly as well. In view of rising energy prices and geopolitical considerations regarding energy imports, it is important to identify and adopt policies and measures that enhance energy security and help reduce the final energy requirements of the economy. An integrated assessment of all the technological options available to the economy is therefore crucial, to examine possible energy pathways and their impacts in terms of costs, infrastructure requirements and fuel mix patterns over time.

This chapter draws on the study 'National Energy Map for India: Technology Vision 2030' carried out by The Energy and Resources Institute (TERI) and funded by the Office of the Principal Scientific Adviser to the Government of India (TERI 2006).

The energy sector in India

In recent years, India's energy consumption has been increasing at one of the fastest rates in the world due to population growth and economic development. Primary commercial energy demand grew at the rate of six per cent between 1981 and 2001 (Planning Commission 2002). Despite the overall increase in energy demand, per capita energy consumption in India is still very low compared to other developing countries – 480 kWh of electricity consumption and 0.50 toe (about 21.3 gigajoules) of primary energy in 2005 (Key Statistics IEA 2007).

Resource augmentation and growth in energy supply has not kept pace with increasing demand and, therefore, India continues to face serious energy shortages. This has led to an increased reliance on imports to meet the energy demand. As per the 2007 World Energy Outlook published by the IEA, India imported 70 per cent of its crude-oil requirements and consumed about 3 per cent of world oil supply. LNG imports account for about 17 per cent of total gas consumption and coal imports for about 12 per cent of the coal supply.

Given India's resource constraints and increasing demand, leveraging technology is key to addressing India's energy security concerns. This chapter, drawing on the National Energy Map study, examines the role that various technological options could play under alternative scenarios of economic growth and development, resource availabilities and technological progress. An integrated modelling framework was used to facilitate the creation and analysis of various scenarios of energy demand and supply at the national level, as well as to provide a detailed representation and analysis at the technological level for each category of resource, as well as sector end-use demand.

As energy demand is driven by GDP and population growth, different GDP growth rates were used to develop alternative scenarios of economic growth for the economy, while a set of population projections from the Population Foundation of India were used to reflect trends in population growth. Population and GDP estimates over the modelling period were used to estimate end-use demand in five sectors of the economy (agriculture, commercial, residential, industrial and transport).

A computer model of the energy system (based on the MARKAL modelling framework) was used to examine the pathways for optimal energy supply to meet end-use services in the five economic sectors under each of the scenarios. Apart from indicating the minimized total system cost of the energy sector under various scenarios, the main outputs provided by the model include information regarding the level of uptake of total energy resources; their distribution across the consuming sectors; the choice of technological options at resource supply, conversion and end use levels; investment levels in each five-year time period; an indication of capacity additions and retirements; and emission levels associated with resource supply and the end-use technological options that are adopted. The modelling time span is from 2001 to 2031 and the data input to the model is from 2001 to 2036. The overall methodology is schematically described in Figure 11.1.

Energy scenarios for sustainable development in India

In order to highlight alternatives, two key scenarios were modelled – the Business-As-Usual (BAU) scenario and the Hybrid (HYB) scenario (both assuming eight per cent GDP growth rate).

BAU scenario

This scenario is characterized by the 'most likely' path of development in the absence of any major new policy interventions, while retaining existing government plans and policies. With regard to technology penetration in the power sector, a limited deployment of clean coal technologies is assumed. The penetration of various renewable energy technologies is modelled as per existing trends and expert opinion. Autonomous efficiency improvements are built into the scenario

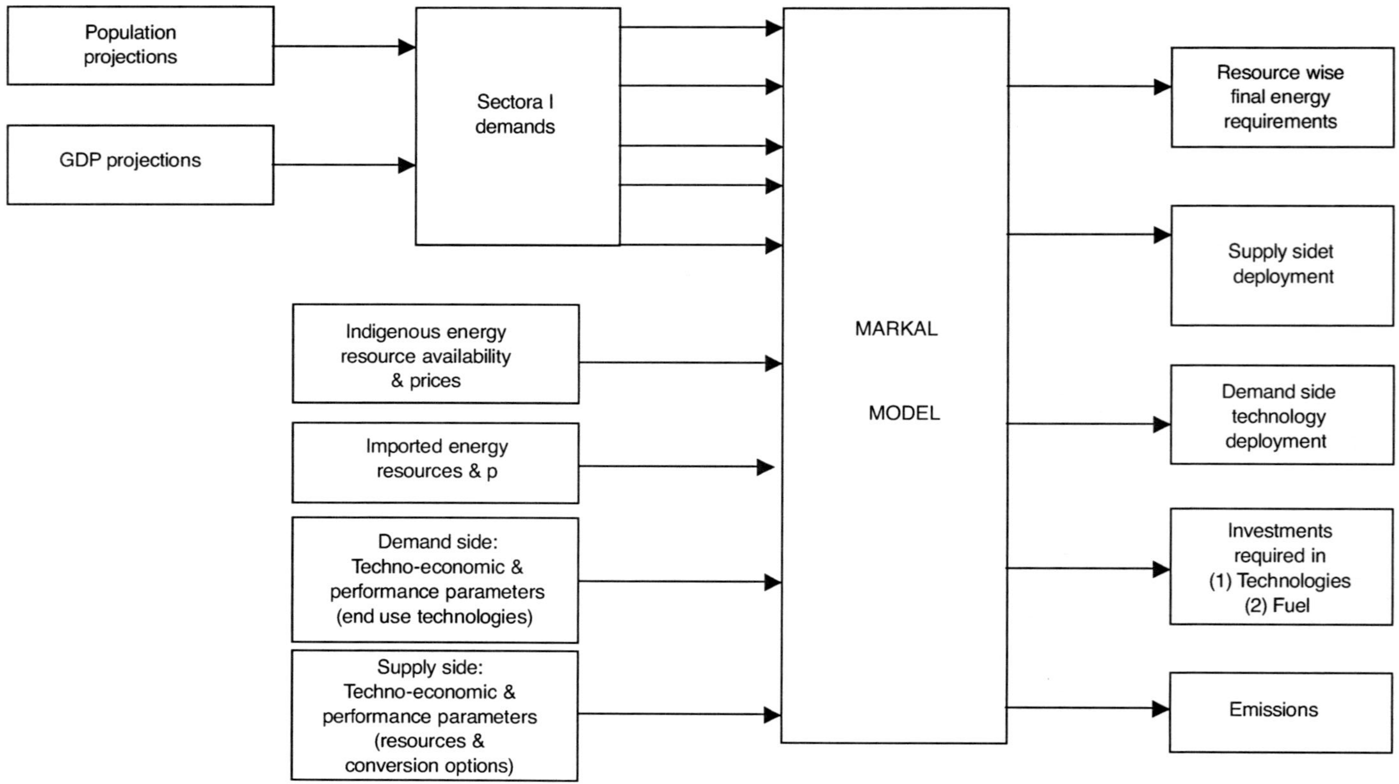

Figure 11.1 Schematic representation of the methodology adopted for the study.

in accordance with technological diffusion and observed efficiency improvements in both conversion and end-use sectors. Thus – although in the long term this is a substantial improvement over the current situation – the BAU scenario still falls short of achieving a transition towards sustainable development.

Hybrid scenario

This scenario is a combination of a number of other scenarios laid out in the National Energy Map document. It adds a stress on renewable energy, increased nuclear energy and a focus on energy efficiency measures to the BAU scenario assumptions. The HYB scenario describes an energy future for the Indian economy involving the incorporation of a wide range of energy-efficiency measures in the end-use sectors, the deployment of clean coal technologies, aggressive penetration of nuclear-based power generation technologies and an aggressive push towards renewable energy sources. In what follows, we describe the various components of the HYB scenario.

High efficiency

This component takes into account energy-efficiency measures spanning the various sectors of the economy. On the supply side, advanced gas-based power generation (e.g. H-frame combined-cycle gas turbines) with 60 per cent efficiency is assumed to be commercially available by the year 2016–17. Renovation and modernization of old coal plants is allowed only up to 2011 as per government plans. In view of the possibility of greater technology transfer across countries and a greater thrust on indigenous R&D in the power sector in this scenario, all clean-coal technologies are allowed to penetrate in an unconstrained manner to their maximum capacity from their year of introduction.

On the demand side, efficiency improvements are considered in various end-use sectors. An increased share of efficient electrical appliances is modelled as meeting the demand for space conditioning, lighting and refrigeration in residential and commercial sectors. In addition, this scenario incorporates a faster rate of displacement of inferior fuels like firewood and kerosene by clean fuels such as LPG for cooking purposes in the residential and commercial sectors. Furthermore, energy-efficiency measures in the transport sector (such as an increased share of rail vis-à-vis road in passenger and freight movement, more public transport and so on), that are driven by policy interventions by the government are also incorporated in this scenario. The industry sector also boasts of measures that lead to significant energy-savings. For instance, in the iron and steel industry, the penetration of efficient blast furnaces and basic oxygen furnaces is allowed up to an 80 per cent level by the year 2036.

Increased nuclear capacity

The installed capacity of nuclear power plants was 2.82 GW in 2001/–02 and 3.31 GW on 31 January 2006[1]. Nuclear-based power generation capacity is expected to increase to 6.78 GW[2] by the year 2010 and further to 21.18 GW by 2020[3] as per the first stage of the Indian nuclear power programme. Beyond 2021, in the BAU scenario, nuclear-based power generation capacity is constrained in the model by the non-availability of indigenous nuclear fuel and by import restrictions caused in part by various geopolitical factors. The 'high-nuclear' scenario assumes importance in view of developments in the nuclear sector with regard to international civil nuclear cooperation and the Government of India's initiatives in this direction. This scenario considers an aggressive pursuit of nuclear-based power generation wherein nuclear-based generation capacity is considered to increase to 40 GW by 2021 and 70 GW by 2031–32, assuming that the country is able to import nuclear fuel (enriched uranium).

Aggressive pursuit of renewable energy

SMALL HYDRO AND WIND

In this scenario a high penetration of renewable energy is considered. The small hydro potential is 10 GW (MNES, 2005a), which is achieved by 2016. Similarly, for wind power generation, the gross potential in India is estimated at 49 GW (MNES 2005a). However, the technically feasible potential is reported at 13 GW (MNES 2005a). In the aggressive pursuit of renewable energy scenario it is assumed that 12 GW of wind capacity could be installed by the year 2036.

SOLAR PV

Average daily solar radiation incident over the land area is in the range of 4–7 kWh/m^2. The potential of solar photovoltaic (SPV) power in India is estimated at 20 MW/km^2 (MNES 2005a). Assuming an aggressive pursuit of renewable energy, the installed capacity of SPV-based power plants is allowed to increase to up to 20 GW in 2036.

BIOMASS

Biomass can be used as a primary fuel by direct combustion, or there can be conversion of raw biomass into secondary fuel (solid, liquid, and gases) by biological or thermochemical conversion processes for thermal applications or power generation. The potential for biomass-based power plants has been estimated to be 16 GW. Current installed capacity is estimated at around 560 MW (with bagasse based cogeneration accounting for another 690 MW). Significant additions are planned by the government in the 11th plan period (2007–2012). The 11th plan addition would include about 1200 MW from bagasse based cogeneration projects and another 500 MW from agro-waste based biomass projects (Pilliai 2007).

BIODIESEL

In addition to power generation technologies, biodiesel is assumed to be available to the transport sector in this scenario. Based on the maximum potential area that is available for plantation for biodiesel production, an upper bound of 31.9 million tonnes of oil equivalent by the year 2031 is imposed on the availability of biodiesel in the aggressive pursuit of renewable energy scenario.

Results and analysis

Having described the two scenarios that will be compared in this chapter, we now move to a discussion of the results obtained from running the MARKAL model in both cases. The differences between the HYB and BAU scenarios are instructive and provide one way to appreciate the opportunities and constraints available to India, as it seeks to define an appropriate energy policy for the future.

Total commercial energy requirements in the BAU scenario

The total commercial energy consumption in the BAU case increases by 7.5 times (6.9 per cent growth rate) over a 30-year period. Table 11.1 presents the commercial energy requirements for fuel. These data are represented pictorially in Figure 11.2.

Coal remains the dominant fuel in the commercial energy mix and the model outputs project an increase in coal consumption from 353 to 2,057 million tonnes during the modelling time frame (2001 to 2031). The major end use of coal remains power generation. However, power generation is projected to account for a declining share of overall coal consumption, dropping from 70 per cent in 2001 to 58 per cent in 2031. This decline is due to an increasing preference for natural gas for power generation, owing largely to better economics. Corresponding to this shift in the fuel mix used to generate power, the share of process heating in coal consumption increases from 14 per cent in 2001 to 24 per cent in 2031.

The percentage share of coal in the commercial energy mix ranges from

Table 11.1 Commercial energy requirements in BAU (mtoe)

	2001/02	*2006/07*	*2011/12*	*2016/17*	*2021/22*	*2026/27*	*2031/32*
Coal	150	193	242	344	466	757	1176
Natural gas	25	36	51	74	132	136	136
Oil	101	151	211	298	405	555	757
Hydro (large & small)	7	9	18	24	30	36	40
Nuclear	2	2	4	8	13	13	13
Renewables	0	1	1	1	1	1	1
Total	285	391	527	749	1046	1497	2123

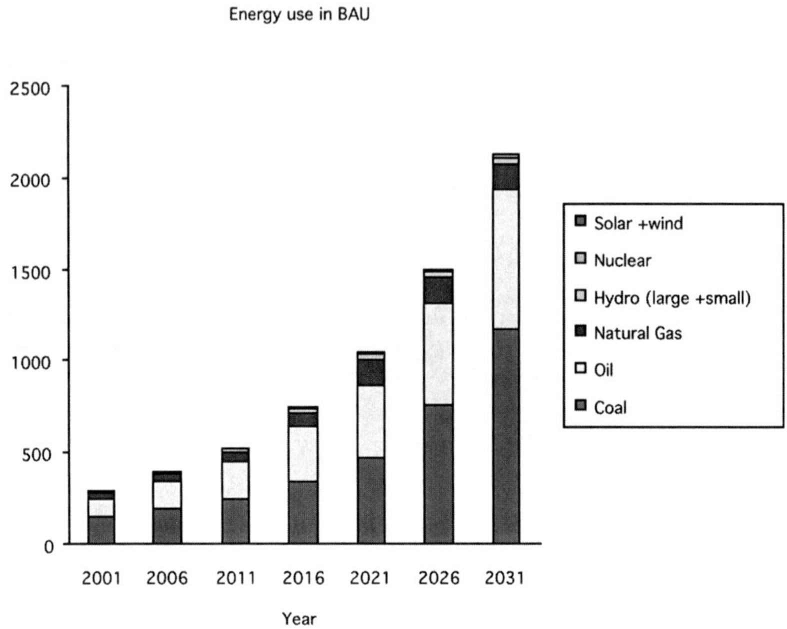

Figure 11.2 Commercial energy use (mtoe) in BAU.

45 per cent to 55 per cent over the entire modelling period. The share of oil in total commercial energy ranges from 36 per cent to 40 per cent during 2001–2031. Oil requirements increase by 7.5 times during the same period.

Although the use of natural gas increases over time in terms of magnitude, its share decreases after 2021 due to non-availability of gas. Indigenous gas production reaches its maximum production capacity by 2011–12 (~44 mtoe). Gas is an economically preferred option for power generation and the fertilizer sectors.

The share of renewable energy (solar, wind, and biodiesel) in commercial energy supply remains less than one per cent throughout the modelling time frame. This is primarily a consequence of the economics of power generation using these options. At this point, it remains necessary for the costs of renewable generation technologies to drop further, if these options are to form a more substantial part of the commercial energy supply. Alternatively, if some of the externalities from fossil fuels were incorporated into their costs, such as through a carbon tax, then renewable energy would become more attractive.

In the BAU case, the use of traditional fuels such as firewood, crop residue and dung in the residential and commercial sectors decreases to half the current level of consumption over three decades. The percentage share of traditional fuels in total primary energy (commercial and non-commercial energy) supplies decreases from 36 per cent in 2001 to four per cent in the year 2031. This is mainly due to a switch from non-commercial fuels to commercial fuels for cooking in the residential sector.

Sector energy consumption in the BAU scenario

Figure 11.3 shows the trend in commercial final energy consumption from 2001 to 2031 in a BAU scenario. The total commercial energy consumption from the end-use side grows 7.5 times during the period 2001 to 2031. The highest contributor towards this growth rate is the rapid increase in oil consumption in the transport sector, which grew by 13.6 times (CAGR of about nine per cent). The rapid growth in the transport sector can be attributed to a shift towards more energy intensive modes of transportation both for passenger and freight movement (enhanced road-based and personalized transportation).

Commercial energy consumption in the industry sector increases by 7.1 times (CAGR of about seven per cent) in the BAU scenario over the period 2001–2031. This rapid growth in energy consumption in the industrial sector is largely on account of the infrastructure requirements of the country (steel and cement demand) as well as small-scale industrial growth.

The overall final energy consumption in the residential sector increases only by 5.3 times from 2001 to 2031. However, during the first two decades, the increase in energy consumption is almost twice that of the base year (2001).

The percentage share of the industrial sector in commercial energy consumption is high throughout the modelling time frame, with this sector accounting for more than 60 per cent of the total commercial energy consumption. Furthermore, the share of the transport sector in total commercial energy consumption is observed to increase from 14 per cent in 2001 to 26 per cent in 2031.

The total consumption of petroleum products increases at the rate of 7 per cent

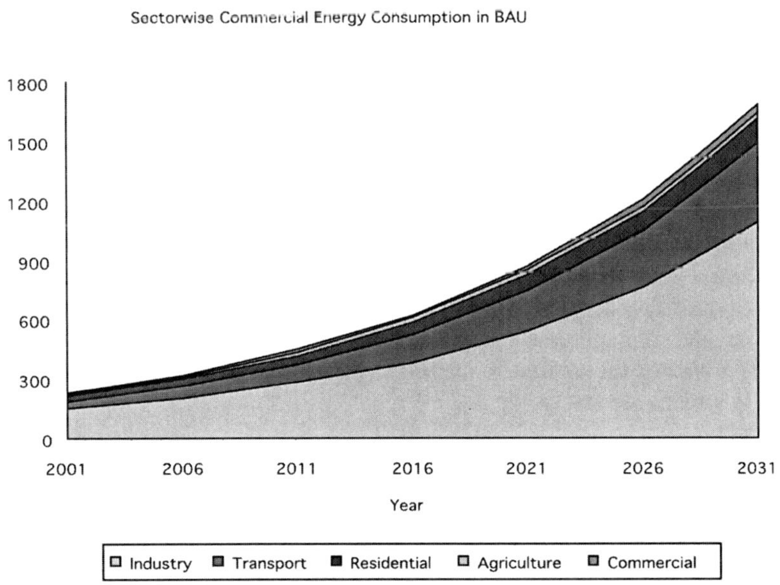

Figure 11.3 Commericial energy consumption (mtoe) across different sectors in BAU.

during 2001–2031, while their consumption in the transport sector grows the fastest at around 9.1 per cent in the same time period. The share of the transport sector in total petroleum product consumption increases from 36 per cent in 2001 to 64 per cent in 2031.

Traditional fuels are used mainly in the residential sector and to a very small extent in the commercial sector. Since these fuels get increasingly replaced with modern energy options such as kerosene and LPG, the net increase in energy consumption in the residential sector is dampened, due to the higher efficiency of commercial energy forms.

The total electricity consumption increases by 8.9 times over the modelling time frame. Electricity consumption increases most rapidly in the industrial and residential sectors, and by 2031, these two sectors account for nearly 80 per cent of the total electricity consumption as compared with 63 per cent in 2001. The electricity consumption in the residential sector increases by 12.6 times from the year 2001 to 2031.

The percentage share of the industrial sector in total electricity consumption in the BAU scenario increases from 42 per cent in 2001 to 51 per cent by the year 2031. During the same period, the percentage share of the residential sector in electricity consumption increases from 21 per cent to 30 per cent. Meanwhile, there is a decline in the percentage share of the agriculture sector in total electricity consumption from 22 per cent to 5 per cent over the modelling time frame. The share of the commercial and transport sectors in electricity consumption has remained constant over the 30-year modelling time frame.

Projected energy balances in the BAU and HYB scenarios in 2031

Table 11.2 and Table 11.3 show the forecast energy balance for the year 2031 under the BAU scenario and the HYB scenario respectively. A comparison of these two scenarios indicates that the final energy requirement in the HYB scenario is about 29 per cent lower than in the BAU scenario. This has been achieved by the adoption of clean coal technologies for power generation, energy-efficient technologies in the end-use sectors, primarily in the industrial and residential sectors, and a major modal shift in the transport sector. A reduction in end-use energy, which is 20 per cent lower in the HYB scenario as against the BAU scenario, also contributes to a reduced fuel demand. The HYB scenario has lower carbon intensity as both the coal and oil requirements are lower, by 409 mtoe and 273 mtoe respectively. There is a marginal increase in nuclear energy and renewable energy, by 29 mtoe and 33 mtoe respectively.

Technology choices in the BAU and HYB scenarios

Power generation technology mix

Figure 11.4 shows the technology-wise power generation capacity mix in the BAU and HYB scenarios for the year 2031. In the HYB scenario, nuclear power

Energy Balance BAU 2031(mtoe)		Coal	Natural Gas	Oil and Petroleum Products	Hydro	Nuclear	Renewables	Total Power	Total
Supply									
		1176	136	757	40	13	1		2123
Conversions									
	Power Generation	215	56	-				325	
Conversions losses & Auxiliary consumption									
	Power generation	448	49						
	Oil Refining			50					
	T&D							68	
Consumption									
	Agriculture	-	-	11				14	25
	Industry	513	31	190				114	848
	Transport	-	-	452				9	461
	Residential	-	-	42				86	129
	Commercial	-	-	12				33	45
	End-use consumption	513	31	708				256	1508

Table 11.2 Projected energy balance for the year 2031 in the BAU scenario

Energy Balance HYB 2031(mtoe)		Coal	Natural Gas	Oil and Petroleum Products	Hydro	Nuclear	Renewables	Total Power	Total
Supply									
		767	136	484	41	42	33		1503
Conversions									
	Power Generation	126	42	-				254	
Conversions losses & Auxiliary consumption									
	Power generation	170	32						
	Oil Refining			41					
	T&D							51	
Consumption									
	Agriculture	-	-	9				10	19
	Industry	471	37	148				86	743
	Transport	-	25	231			28	17	302
	Residential	-	-	42				65	107
	Commercial	-	-	13				25	38
	End-use consumption	471	62	444				204	1209

Table 11.3 Projected energy balance for the year 2031 in the HYB scenario

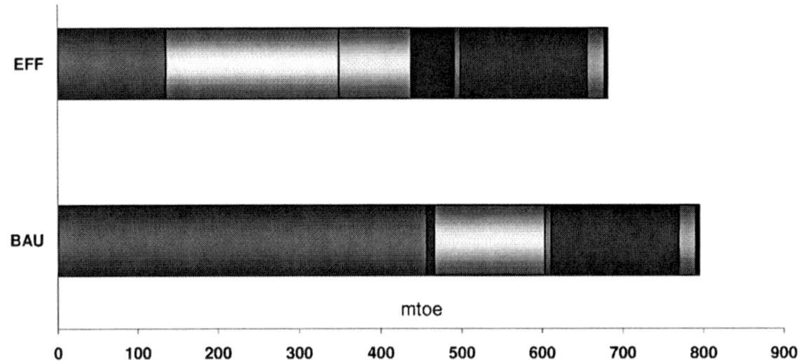

	BAU	EFF
Coal Subcritical	456.36	134.82
Coal Efficient	9.58	0.5
Coal IGCC	0	213.3
Gas	137.4	88.6
Gas Efficient	0	52.84
Diesel	7.97	7.97
Hydro	158	158
Nuclear	21.18	21.18

Figure 11.4 Power generation technology deployment in the BAU and HYB scenarios for 2031.

generation displaces coal based power generation and gas based power generation and hydro based power generation capacity increase marginally. Renewable energy generation capacity reaches its maximum potential of 26 GW in the HYB scenario in the year 2031. Hydro based power generation and nuclear based power generation are exploited to the maximum potential of 160 GW and 70 GW respectively in the year 2031. However, coal-based generation capacity still accounts for about 42 per cent of the total power generation capacity in the HYB scenario.

Figure 11.5[4] presents the trends of energy intensity over the modelling time-frame resulting from such technology choices.

Figure 11.5 clearly depicts that energy intensity exhibits a declining trend from 0.022 kgoe/Rs. of GDP in 2001 to 0.017 kgoe/Rs. of GDP in 2031 (a decrease of 23 per cent) in BAU. It can be inferred that even in the BAU scenario based on a GDP growth rate of 8 per cent and incorporating government plans and policies, the economy is progressing along a path of improving efficiency. However, the BAU scenario takes a conservative view with respect to technology deployment, with limited penetration of clean-coal technologies, H-frame combined cycle gas turbines, slow penetration of efficient power-generation technologies, and a low degree of penetration of nuclear power and renewables. In the HYB scenario (which includes all plausible energy-efficiency measures considered in the high-efficiency scenario coupled with enhanced nuclear capacity and accelerated penetration of renewables), the energy intensity steadily declines from 0.022 kgoe/Rs. of GDP in 2001 to 0.012 kgoe/Rs. of GDP in 2031. The energy intensity in this scenario is around 29 per cent lower than the corresponding level in BAU for the year 2031.

To use the residential sector as an example – under a BAU scenario, the total

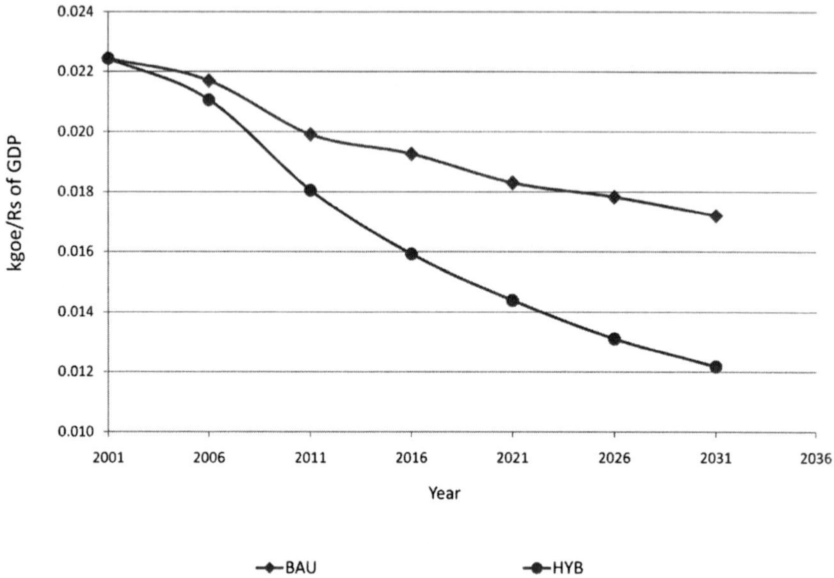

Figure 11.5 Trends in energy intensity from 2001–2031.

commercial energy consumption in this sector increases from 25 mtoe in 2001 to 129 mtoe in 2031 at a CAGR of 5.6 per cent. The consumption of petroleum products increases from 17 mtoe to 39 mtoe during the same period. However, electricity consumption grows at a much faster rate, from 82 TWh (in 2001) to 1,034 TWh (in 2031) (12.6 times). This is due to the Government of India's policy to provide electricity to all by 2012 and also due to changing lifestyles and increased use of electrical appliances for cooking, food preservation and space conditioning. The demand for electricity is mitigated in the HYB scenario, where 2031 consumption is projected as being only 768 TWh, a result indicative in part of the increased use of energy-efficient appliances in that scenario.

The modelling results thus suggest that there exists considerable scope for bringing about reduction in energy intensity in India, if policies are formulated to promote clean-coal technologies (in view of the economy's continuous dependence on coal) and if there is removal of barriers to the uptake of more energy-efficient technology options. In this direction, with time-bound targets and concerted action plans towards strengthening indigenous R&D facilities, there is the possibility of further reduction in energy intensity, as highlighted in the HYB scenario. There is room therefore to both significantly reduce the country's demand for power-generation capacity and to change the way in which that demand is met.

Energy consumption in various transport scenarios

The other key sector of concern in India is the transport sector. To examine the options for this sector, the model used various scenarios, shown in Table 11.4.

Table 11.4 Scenarios in the transport sector

Scenario	*Description*
Enhanced share of public transport (PUB-PVT)	Share of public transport increased to 60% in 2036 as against 51% in the BAU scenario
Increases share of rail in passenger and freight movement vis-à-vis road (RAIL-ROAD)	• Railway freight share increased from 37% in 2001 to 50% in 2036 as against 17% in the BAU scenario • Railway passenger share increased form 23% in 2001 to 35% in 2036 as against 23% in the BAU scenario • Share of electric traction increased for rail passenger and freight to 80% by 2036 instead of 60% in the BAU scenario
Fuel efficiency improvements (FUEL EFF)	Fuel efficiency of all existing motorized transport modes increase by 50% from 2001 till 2036
Enhanced use of bio-diesel in transport sector (BIO-DSL)	Penetration of bio-diesel to 65 Mtoe by 2036
Transport sector hybrid (TPT-HYB)	Incorporates all the above-mentioned measures in addition to BAU

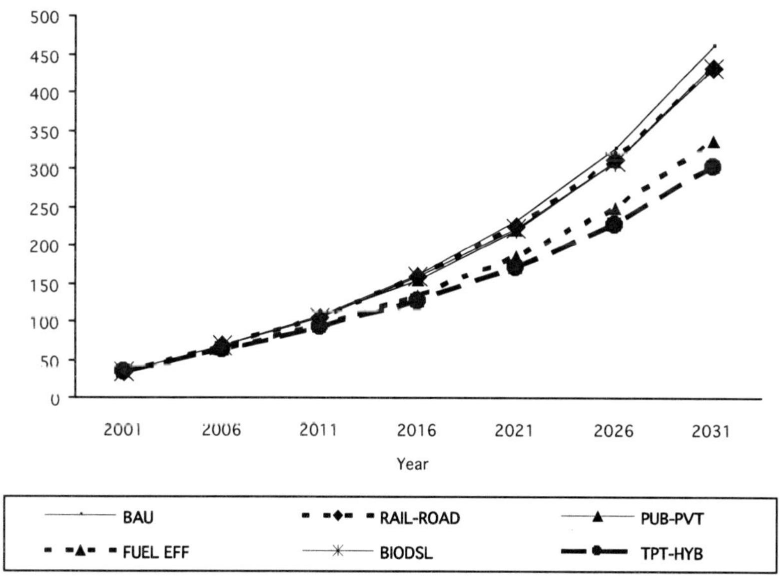

Figure 11.6 Comparison of energy consumption (mtoe) in the transport sector across scenarios.

Figure 11.6 provides a comparison of total commercial energy consumption (including electricity) in the transport sector across various scenarios. Figure 11.6 clearly indicates that the projected energy consumption (including electricity) in the transport sector exhibits a consistently upward sloping trend in all the five scenarios (as well as in the BAU) over the 30-year period (2001–2031). In all transport sector scenarios, freight and passenger transport demand shows an upward moving trend. The figures for projected energy consumption in the transport sector are based on the optimal fuel-technology mix in the transport sector in each scenario.

Total energy consumption in the transport sector increases from 34 mtoe in 2001 to 461 mtoe in 2031, registering an average annual growth rate of 9.1 per cent during the period 2001–2031. However, as shown in Figure 11.6, it is possible to achieve reductions in energy consumption up to a maximum level of around 35 per cent in 2031 in the transport HYB scenario vis-à-vis the BAU scenario. In absolute terms, energy consumption in the HYB scenario declines by 125 mtoe for the year 2031 vis-à-vis the BAU scenario.

Table 11.5 Suggested technology deployment programme

now to 2011	*2011 to 2021*	*2021 to 2031*
Power Generating Technologies		
Deploy • Hydro power generation • Super critical boilers/ ultra super critical boilers Advanced gas turbines ex: H-frame turbine Refinery residue based IGCC Demonstration of commercial scale IGCC plants using indigenous and imported coals Fast breeder nuclear reactor	Commercialize IGCC Ultra super critical boiler commercialized	Demonstration of commercial scale Thorium based reactors demonstrated.
End use technologies		
Cogeneration Use of waste recovery in industrial processes Lighting technologies: CFL, LED, Energy efficient white goods: refrigerators, AC T&D loss reduction: HVDC, HVAC & amorphous core transformer	Adopt state-of-the-art industrial processes	Adopt state-of-the-art industrial processes
R&D in exploration and production of fuels		
Natural gas from gas hydrates in-situ coal gasification Deep sea natural gas Coal bed methane (CBM) Mining of coal from seams greater than 300 meters	Commercialize • In-situ coal gasification • Deep sea natural gas • CBM • Mining of coal from seams greater than 300m	Natural gas from gas hydrates

Conclusion and technology pathways

The model results indicate that the greatest reductions in energy consumption in India can be achieved by carrying out interventions in the power sector on the supply side; and in the transport and residential sectors on the end-use side.

The model results also indicate that if all power generating technologies were allowed to compete for new capacity additions, the preferred choice of technologies in the order of economic merit would be (1) Large hydro; (2) Refinery residue based IGCC; (3) imported coal-based IGCC; (4) high efficiency CCGT (H-frame gas turbine); (5) indigenous coal-based IGCC; (6) normal CCGT; (7) ultra super critical boiler; (8) super critical boiler.

The analyses of the model results at the end-use side indicate that maximum impact on final energy demand can be achieved by the adoption of energy-efficient technologies in end-use sectors like the transport and residential sectors and by adopting state-of-the-art industrial-process technologies. This is in consonance with similar conclusions in other parts of the world (see Rosenfeld (2008) for a discussion of the importance of energy efficiency in a US context) and underscores the importance of energy efficiency measures as a means of tackling our energy security and climate concerns.

In addition, the results indicate that India's commercial energy demand will grow 7.5 times over the period 2001–2031. Further, India's dependency on energy imports will increase significantly over these 30 years, with coal-import dependency expected to increase from 3 per cent to 70 per cent and oil-import dependency increasing from 68 per cent to 90 per cent during the same period. It becomes imperative to increase the supply of indigenous energy resources and aggressively adopt energy-efficient technologies across all sectors of the economy. India should therefore plan to enhance efforts in R&D in the exploration and production of energy resources – especially in the area of deep sea natural gas exploration; technologies to exploit coal from seams which are more than 300 metres deep; in-situ coal gasification; and gas hydrates.

A brief overview of the status of various technologies and recommendations for their deployment are given in Table 11.5.

Notes

1 Ministry of Power, Government of India.
2 Nu Power, Vol. 18 (2–3), Department of Atomic Energy, 2004.
3 Anil Kakodkar, Department of Atomic Energy.
4 The energy intensity indicates the extent to which energy is efficiently utilized in generating a unit of income/output (GDP) for the economy.

References

IEA 2007. Key World Energy Statistics 2007. Paris: OECD/IEA. pp. 53

Pillai, V., 2007. 1700 MW biomass power capacity in 11th plan. *Project Monitor*. December 10–16, 2007. Available at <http://www.projectsmonitor.com/detailnews.asp?newsid=15049>. Accessed 15 September, 2008.

Planning Commission. 2002. *Report of the Committee on India Vision 2020*. New Delhi: Planning Commission, Government of India.

Rosenfeld, A., 2008. 'Energy efficiency: the first and most profitable way to delay climate change.' Presentation by Commissioner Art Rosenfeld to the meeting of the U.S. Environmental Protection Agency, Pacific Southwest Region IX, in San Francisco, California, 9 June, 2008. Publication # CEC-999-2008-013. Available at <http://www.energy.ca.gov/2008publications/CEC-999-2008-013/CEC-999-2008-013.PDF>.

TERI, 2006. 'National energy map for India: technology vision 2030.' Report No: PSA/2006/3, Office of the Principal Scientific Advisor, Government of India, New Delhi: TERI Press.

12 Incentivizing change in energy choices

Ajay Mathur, Koshy Cherail, and Deepti Mahajan

The challenge of meeting the world's energy needs while mitigating climate impacts, and adapting to them, has emerged as a priority area in domestic and international policy. There is an urgent need to ensure access to clean and adequate energy to all people in the world, and to enable continued economic development. At the same time there exists a global imperative to stabilize greenhouse gas emissions. Achieving these twin goals will require both reaching a global compact on emissions' allocations, and accelerating technological change so as to decouple energy intensity and development. Technology is probably the key to enabling a future where energy service needs, energy security concerns and climate constraints are addressed simultaneously. Through the use of appropriate energy sources and technologies, it is possible to avert the worst consequences of climate change while expanding energy supplies to meet the needs of both developed and developing countries. An increase in the energy efficiency of end-use technologies, accompanied by a gradual switch to renewables and low-carbon energy sources, is crucial to fulfilling these dual objectives.

Even as developing countries negotiate an equitable global emissions' allocation regime, many countries, including India, are devising policies and implementing actions to secure their energy future, and to move towards a low-carbon economy as a co-benefit. A transition to clean and energy-efficient technologies for consumer and industrial applications is a key component of India's sustainable growth trajectory. This transition includes a focus on enhancing energy efficiency in the power, industry, and household sectors, and in commercial and residential buildings; promoting the use of efficient appliances; facilitating a shift to renewables (biomass, wind, solar energy) and other low-carbon electricity-generation technologies; and promoting the use of efficient vehicles and greater use of public transport. The decisions made in the next five years will be crucial, as they will lock in the future path of technology development and diffusion, and resource exploitation, for the bulk of India's eventual infrastructural investments.

Clearly, technology innovation and adoption requires not only a focus on research and development, but also political mobilization, policy and regulatory changes as well as changes in institutional structures. This chapter provides an overview of the forces that hinder the adoption of available technologies for enhancing energy efficiency and reducing dependence on conventional fuels. We also look at some policy interventions required to reduce or eliminate these forces, to and create incentives and pressures for the adoption and deployment of such technology.

Technological possibilities: unscrambling the energy intensity–economic growth linkage

The need to manage energy consumption in an environment-friendly manner while maintaining high growth rates calls for a proactive delinking of economic growth from increasing energy use. Since developing countries cannot quickly move away from the development path they have adopted without adversely impacting growth rates, policymakers worldwide have now begun to recognize that if climate change is to be effectively addressed, developmental and growth-related issues need to be integrated into any framework for mitigation and adaptation. At a national level, however, India has taken a lead in adapting its technology infrastructure and policymaking process to address the challenges of energy insecurity and climate change. Development goals have aimed at balancing economic growth and environmental concerns. Reforms in the energy sector have accelerated economic growth and enhanced the efficiency of energy use. India has achieved some success in loosening the linkage between energy and GDP growth at an early stage of development. The country's energy intensity of GDP has reduced from 0.30 kgoe per dollar GDP in PPP terms in 1972 to 0.19 kgoe per dollar GDP in PPP terms in 2003; this is equal to that of Germany.

Technological change has been an important component of the strategy to effect change in energy-use patterns. Over the past decade, energy efficiency in Indian industry has increased steadily, due to competitive pressures and rising fuel prices. In all the major energy intensive sectors – steel, aluminium, fertilizer, paper, and cement, levels of energy efficiency in new plants meet the prevalent global levels. In the cement sector in particular, the energy efficiency of Indian plants is amongst the highest in the world. In these major energy-consuming industrial sectors, average specific energy consumption has been declining, partly because of energy conservation in existing units, and to a much larger extent thanks to the addition of new capacity using state-of-the-art technology. Several other initiatives to promote energy security and energy access are also being put in place, which move the country towards a low-emissions growth path. These include increasing the capacity of renewable energy installations; improving the air quality in major cities (by introducing CNG-fuelled public transport); promoting the energy efficiency of consumer appliances, and initiating afforestation programmes. India has also directly contributed to greenhouse gas (GHG) mitigation through the CDM mechanism, and hosts the largest number of registered projects among all countries (UNFCCC 2008). These have contributed to an effective delinking of the energy sector's growth from economic growth. Currently, primary energy sector growth rate is 2.76 per cent per year, as against a GDP growth rate exceeding 8 per cent.

India also has a multifaceted renewable energy programme supported by a dedicated ministry, the Ministry of New and Renewable Energy. The renewable energy sector in India is dominated by the private sector, which accounts for about 95 per cent of the total investment in the sector. Government policy has created a renewable-energy market, and helped direct these investments into profitable investments. Numerous fiscal and financial incentives have been instituted for the promotion and exploitation of renewable energy, with the Indian Renewable Energy Development Agency Limited (IREDA) serving as a dedicated financial institution for renewable energy.[1] The sector has a strong R&D orientation, whereby India is emerging as an important manufacturer and exporter of renewable energy technologies such as wind-energy generators, biomass gasifiers and solar PV systems.

That said, even while the country has taken some important steps, a sustained momentum is required to ensure an energy-secure future. The deployment of certain technology options is beneficial not only from the point of view of enhancing the energy security of the nation, but also in terms of achieving increased cost effectiveness and efficiency for the consumer and industry. Unfortunately this win-win situation for all stakeholders is not reflected in the rate of adoption of these technologies, and in many cases they remain underutilized with low market penetration. The following section deals with this gap in the adoption of effective technology.

Challenges and possibilities

It is important to assess the reasons capable energy technologies are not being picked up. What are the market failures that hamper technology acquisition and use? What is required to influence consumer choices in favour of green technologies? What policy and regulatory instruments are required to promote their spread? Addressing these issues calls for a multi-pronged, concerted effort that encompasses awareness creation; attitudinal change; technological R&D; institutional changes; and innovative fiscal and policy measures. A global report by McKinsey & Company (2007) suggests that targeted policies to remove distortions and overcome market imperfections can help capture the opportunities that are available to improve energy productivity and reduce energy demand growth, a first step to climate change mitigation. These policy options to accelerate the pace of technology innovation and up-take include tightening fuel-economy standards, mandating standards for stand-by power consumption[2] and introducing CFLs in a phased manner.

Specific policies for commercial buildings include enforcing building codes, introducing office equipment standards, and labelling. Innovative approaches by utilities, power companies, and ESCOs can help consumers make more informed energy choices and benefit from short pay-back energy-saving opportunities. Of course utilities need to be incentivized to run such programmes. Governments can also encourage higher energy productivity through pilot projects and energy audits, as well as consider providing subsidies or tax credits to companies that are implementing select energy conservation technologies.

It must be noted that requiring technology to operate at various levels is context specific: what may work for one sector may not apply to another; and what may work for another country may not be applicable to India. The typology of technology applications comes with attendant barriers and thus requires specific solutions.

Consumer technologies

Currently, the residential sector in India consumes a significant proportion of the country's total energy,[3] and this is expected to grow rapidly as incomes rise and households seek access to clean and adequate energy supplies. A change in energy-intensive lifestyles can lead to significant reductions in energy use and emissions. The use of energy-efficient technologies for domestic purposes can contribute significantly towards mitigating an increase in energy use. The Indian market today makes available efficient air conditioners, refrigerators and tube lights. Solar lighting systems and water heaters offer possibilities for the use of renewable energy for domestic consumption, and miniaturization of technology will only facilitate the use of efficient embedded generation technologies at the level of the household (Jamasb *et al.* 2006). However, awareness of efficient appliances and renewable energy technologies is low. Often, consumers are unaware of the availability of these technologies. At other times they lack information on the benefits that accrue from their use, and are therefore unable to make an informed buying decision. Most consumers do not see the higher buying price of an efficient technology in light of the pay-offs offered by reduced electricity bills and efficient use of energy resources. In the case of renewable energy technologies, the unavailability of reputed installers and service providers may add to the consumer's inhibitions.

To aid the consumer in making a decision, an Energy Labelling and Certification Programme for appliances was launched by India's Bureau for Energy Efficiency[4] in 2006, wherein a comparative star-based labelling system has been introduced for fluorescent tube lights, refrigerators, air conditioners, motors, and distribution transformers. Figure 12.1 shows the labels introduced for refrigerators and tube lights. The labels provide information about the energy consumption of the appliance, enabling consumers to make an informed decision. Almost all fluorescent tube lights sold in India, and about two-thirds of the refrigerators and air conditioners, are now covered by the labelling programme.

A progressive approach towards promoting efficient consumer appliances should focus on consumer empowerment to influence choices about energy consumption. The certification of goods and labelling for comparative assessment of efficiency should be part of a larger consumer education plan, possibly a well-targeted government initiative to spread awareness about efficient energy use. The move towards efficient consumer technologies also necessarily includes plugging the information gaps that currently mark the manufacturer-consumer relationship. A mediation mechanism that conveys accurate information to consumers about products, and provides consumer feedback to manufacturing companies, needs to be put into place.

In addition, a mechanism is required to buffer the high initial cost of an efficient

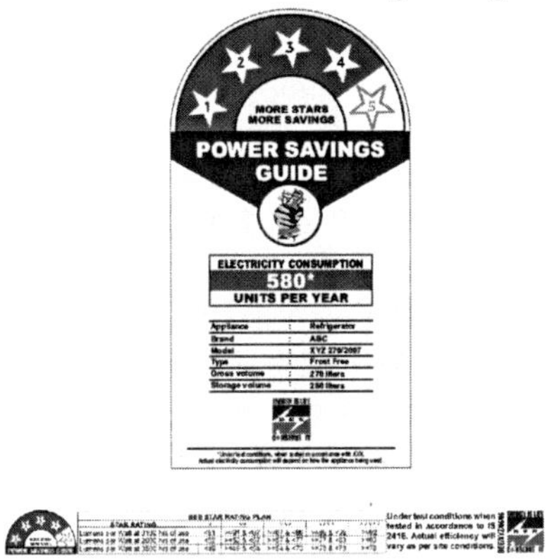

Figure 12.1 BEE energy labels for refrigerators (top) and tube lights (bottom).

appliance/renewable energy technology. Since buying price and maintenance costs are vital to a consumer's decision, high price is seen as a primary deterrent to efficient technologies being bought in the market. Public procurement policies may offer help in this regard, as the bulk purchase of technologies can play an important role in reducing initial costs, thus facilitating their entry into consumers' homes.[5]

Industry technologies

For industry, energy price volatility, energy scarcity, and global climate change are sources of risk and opportunity to be understood and managed. In this context, the drivers of change may range from changing consumer values, and cost and regulatory imperatives, to the demands of competition. Public concern about the environmental consequences of industrial growth could strengthen market pressures favouring 'green' companies. It may affect a company's ability to market their products and mobilize investment for industry processes if they are perceived as 'dirty.' Moreover, due to developments in markets, available knowledge and technology, new companies may take environmental leadership, putting pressure on competing units to 'green' themselves. For large-scale industries, efficient use of energy and use of alternative fuels and renewable energy sources can also lead to cost savings and competitive advantages while facilitating upgrading of the manufacturing process.

In light of the changes mentioned above, businesses have begun to take a lead in responding to the threat of climate change and energy insecurity. Globally, there are several examples of such efforts, including the formation of the Business Council for Sustainable Development (BCSD); setting of voluntary emission-reduction targets by companies (primarily traded in CCX); and corporate investment in clean technology R&D. Companies across the world are improving the efficiency of their business processes, and reducing energy and material consumption by utilizing instruments such as supply-chain management, enterprise resource planning (ERP), and automation. Supported by the government, Indian industry too has demonstrated leadership in the area. Indian companies have been deploying cutting-edge technology, while building new plants and refurbishing old ones. They have been pursuing acquisitions abroad, mainly to gain access to technology. The Green Business Centre, set up under the leadership of the Confederation of Indian Industry (CII), has become a hub for promoting environment-friendly business practices. The CII Climate Change Centre was established in 1998 to advise members on emerging challenges. Industry has also facilitated CDM projects in the country, and has undertaken voluntary measures such as carbon disclosure.

Further, much of the investment for renewables-based installed capacity (which stands above 11,000 MW currently), has come from the private sector. India has already established a robust renewable-energy manufacturing and R&D base: companies like Signet and Moser Baer are setting up solar manufacturing units to add to the existing capacity of Tata BP, CEL and so on. Companies such as Suzlon, Enercon, and Vestas India have set up large wind-energy generator manufacturing units.

Yet, if technological change is to be pursued on a large, accelerated scale, concerted technical and policy support is required. In the absence of performance guarantees, businesses may find it difficult to invest large sums of money in new technologies. The element of uncertainty over expected benefits and an inability to measure savings become an impediment to the introduction of efficient, environment-friendly technologies and processes in the industrial sector. Industry players therefore need to be assured of technology performance and given incentives to introduce desirable changes in their technology infrastructure. Guarantee and insurance schemes can play an important role in addressing these concerns. What is also significant is that technology manufacturers take into account industry feedback when upgrading products and services. Adaptive technology development that caters to the specific requirements of particular industrial sectors will clearly have more takers in the industry.

Energy service companies (ESCOs) are also being promoted so as to better manage and share risks in the adoption of new technologies. Performance contracting by ESCOs enables both technical aid and funding for energy efficiency investments. An ESCO is able to deliver performance-based energy and environmental solutions (technical and financial services) to achieve planned cost reductions through defined risk sharing. While offering a streamlined approach to facility improvements, the ESCO undertakes the technology risk involved, and minimizes or eliminates the problem of up-front cash flow by recovering costs through savings shared with the client (Vankani 2005).

In India, the Energy Conservation Act of 2001 establishes a legal framework to promote the efficient use of energy in all sectors of the economy. The Act empowers the government to direct industry to appoint energy managers; conduct regular energy audits; comply with specific energy consumption norms; and submit reports to the government. The Bureau of Energy Efficiency (BEE) is in the process of fixing norms for specific energy consumption in the cement and the paper and pulp industries, and has set up taskforces for five other energy-intensive sectors. In March 2007 the conduct of energy audits was made mandatory in large energy-consuming units in nine industrial sectors. These units, notified as 'designated consumers' are also required to employ 'certified energy managers', and report energy consumption and energy conservation data annually.

This policy environment takes energy efficiency from being a boiler-room issue to a board-room issue, and creates a market for energy-efficient technologies and services.

Efficient electricity production

In the electricity sector, while it is important to focus on demand-side measures, supply-side efficiency is also important. Clearly, deployment of appropriate, viable technologies is a primary input for efficient production of electricity from conventional and renewable sources for grid-based electricity, and efficient decentralized and distributed generation based on renewables. Even though the Ministry of Environment and Forests in India mandates reduction of air, water and solid waste pollution from power plants, it does not provide guidelines for overall plant efficiency (Chikkatur *et al.* 2007). A few state electricity regulatory commissions and the central electricity regulatory commission have linked the power-sales tariffs from existing thermal power stations to energy-efficiency increases. In addition, the Government of India has notified a tariff policy based on lowest-cost (and hence most energy-efficient) bidding, and is requiring that the new ultra mega power projects be based on the efficient super-critical power-generation technology. Regulatory interventions are thus essential to fuel energy efficiency in electricity generation.

Effective regulation must attack the constraints to the adoption of efficient, cutting-edge technology in electricity generation units, which include, most importantly, the issue of high costs in the short term. Easy availability of finance can reduce risks to some extent but the return on investment will clearly depend on how costing of electricity is done – taking discount rates into account and the time period for which tariffs are determined. The recent restructuring of the electricity sector in India offers an important route to improving power plant efficiency, through regulatory mechanisms that allow for an independent tariff-setting process for bulk purchases of electricity from generators. Tariff-based incentives for efficiency improvements will eventually benefit consumers by reducing electricity costs (Chikkatur *et al.* 2007).

To promote the use of renewables for electricity generation, the government has already taken significant steps. Section 86 (1) (e) of the Electricity Act 2003 requires each state's regulatory commission to specify the percentage of electricity that the distribution companies must procure from renewable sources. Many regulatory commissions have issued orders for Renewable Portfolio Standards (RPSs), and notified preferential prices for electricity from renewables. This has supported the acceleration in renewable-electricity capacity addition, and, over the past three years, about 2,000 MW of renewable-electricity capacity has been added in India every year, bringing the total installed renewable capacity to over 11,000 MW. Of this, a little over 7,000 MW is based on wind power, making India the country with the fourth largest installed wind power capacity in the world. The National Hydro Energy Policy has resulted in the accelerated addition of hydropower in India, which now stands at over 35,000 MW.

Another avenue for the exploitation of renewable energy sources is offered by the decentralized and distributed generation of power, being seen today as a sustainable method to provide access to electricity to rural areas where traditional power lines have not penetrated. Decentralized and distributed generation involves the installation and operation of small modular power-generating technologies primarily based on local renewable energy sources. Modern electronics have also made it possible to integrate these technologies with grid-based large power plants. The Working Group on Power for the Eleventh Five Year Plan has estimated a potential addition of 10,000 to 15,000 MW capacity through distributed generation in the Eleventh and Twelfth Plans.[6] Biomass gasifiers, solar photovoltaic systems, and wind-electric generators are some of the commonly used distributed generation systems for rural electricity supply and distribution. The spread of these distributed generation units, tapping renewable energy, requires access to appropriate renewable energy technologies and formulation of effective business modules. Government regulation in this regard must therefore focus on instituting standardization norms for distributed generation systems, building technical capacity for ensuring optimal performance of field-based systems, and managing and regulating the price of technology through bunch procurement.

Efficiency in buildings

At the national level, domestic and commercial buildings account for about 30 per cent of annual electricity consumption. This consumption figure is slated to rise as India's real estate market grows, and as the growth of the services sector creates demands for commercial office space, even while large residential complexes, shopping centres, malls, and hotels continue to mushroom. Apart from energy-intensive construction activities, modern buildings have high levels of energy consumption because of the requirements of air conditioning and lighting. Energy consumption in the building sector can be reduced by sustainable building design, efficient lighting and space conditioning, and use of renewable energy technologies. Most commercial buildings in India have an energy performance index (EPI) of 200 to 400 kWh per sq. metre per year. Energy-conscious building design has been shown to reduce the EPI to 100 to 150 kWh per sq. meter per year in India, but such initiatives have so far been restricted to environment-sensitive corporate groups (Mathur 2007). According to a study conducted by The Energy and Resources Institute (TERI), in new buildings, such measures offer 40–50 per cent energy saving potential. Retrofit options with efficient lighting, air conditioning and electrical systems offer a 20–30 per cent saving potential.

Sustainable building design involves environment-friendly decision making right from the stage of site selection and planning, and includes the use of energy-efficient construction technologies, renewable energy technologies (where possible), efficient lighting and space conditioning systems, and technologies for the recycling and re-use of water and waste. The incremental costs incurred from implementation of energy efficiency measures in buildings (with respect to the overall cost of projects) vary from 10 per cent to 15 per cent for commercial buildings. Lifecycle cost analyses of the initial investment and operation costs indicate that the payback period is approximately five years.

With a short payback period, sustainable building design offers the possibility of large long-term cost savings. In spite of this, sustainable building design has not become an integral part of building construction in India. The disincentives are manifold. Large scale energy-efficient building design is limited due to the prevalence of split incentives: builders fear that they would bear the costs, while tenants would enjoy the benefits offered by sustainable design. Often buyers, sellers, and builders as well as architects are unaware of the possibilities offered by green building design elements. At other times, consultants are not able to inspire confidence in clients, resulting in inertia. If additional capital expenditure is required and the client is not convinced of the resulting benefits, the offered technologies and services cannot be expected to take off (Shahi 2006).

It is pertinent in such a situation for regulations to put in place a common baseline enforcing minimum performance standards – an effective tool to address split incentives. The Energy Conservation Building Code (ECBC), introduced for voluntary compliance in May 2007, serves this purpose. The ECBC sets minimum energy efficiency standards for design and construction. It encourages energy-efficient design of new buildings and the retrofit of old structures such that the construction design and technology do not constrain the functions of the building; do not adversely influence the health, comfort and productivity of the occupants; and minimize lifecycle costs (construction plus energy costs). It covers guidelines for the building envelope (walls, roofs, windows); lighting; heating, ventilation and air conditioning; solar water heating; and other electrical systems. The application of the Code, as shown in Figure 12.2, reduces energy consumption by about 50 per cent.

Nearly one hundred buildings are already following the Code, and compliance with it has been incorporated into the Environmental Impact Assessment requirements for large buildings. The code is slated to become mandatory after providing for time for capacity building and some initial experience in implementation. To meet this objective, capacity building and training for architects, builders and consultants is urgently required. The availability of information, technical expertise and services can go a long way in altering the energy landscape of the construction business.

Energy conservation retrofits in existing buildings are also being undertaken under an Energy Efficient Government Buildings programme. Innovative financial instruments to promote performance contracting are being developed to overcome the lack of efficient delivery mechanisms (Mathur 2007). In particular, government buildings are being encouraged to upgrade energy use through agreements with Energy Service Companies (ESCOs) which invest in the energy upgrades, and are paid through the savings in the energy bill. This programme has helped to create confidence in the ESCO business model, both for ESCOs and for building owners,

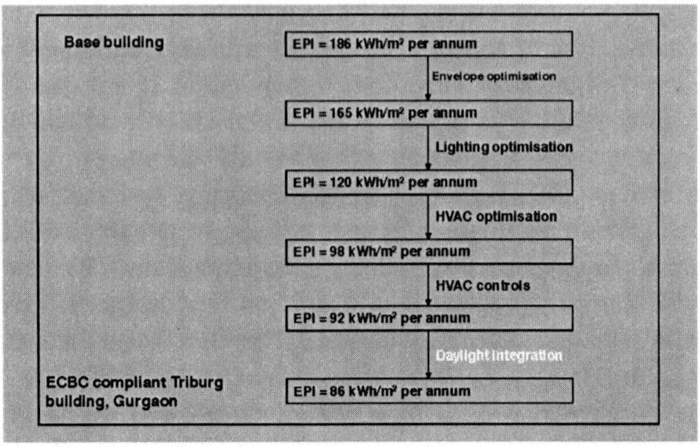

Figure 12.2 Application of ECBC reduced the energy demand by more than 50 per cent in a new building being constructed near Delhi.

and is helping to seed the market for improvements in building energy efficiency improvements through performance contracting.

Consolidating policy to effect change

Any policy on methods to achieve energy security and mitigate climate change needs to be in consonance with the economic and social realities of the country. Involving all stakeholders in decision-making is key to the successful implementation of policies. Because policy options need to be implemented using a multi-actor, multi-level process, it is essential to establish effective institutions that are able to provide a structure for participation and are able to rally stakeholder participation and create lobbies for desired policy changes. While the BEE has furthered policy and regulation for energy efficiency, smaller support organizations need to be established to directly interact with the end user and offer technical consultancy and services. Also, to mitigate the perception of financial and technological risks and barriers, the government needs to step in to support investments in pilot projects that serve as models for the introduction of new technologies within the given political, social and economic context.

As discussed, several factors hamper decision making on energy and technology choices, including lack of awareness about energy efficiency and renewable energy opportunities; lack of access to capital; misplaced incentives, which separate responsibilities for making capital investments and paying operating costs; transaction costs; and bounded rationality (Sathaye and Murtishaw 2004). Government interventions to reduce information and transaction costs can significantly increase the uptake of resource-efficient products and renewable sources of energy. The use of minimum performance standards offers a powerful policy instrument, along with concerted efforts to spread the message, and technical and financial support where required (Sathaye and Murtishaw 2004).

Technology, finance, regulation and policy must work in consonance to enable the transition of the country to a low-emission, energy-secure economy. At the international level, an agreement is required amongst different countries on collaborative strategies for the mitigation of climate change, and for the investment of trillions of dollars (which in any event will be invested in energy in the coming decades) in areas that ensure the world's future needs are met in a safe and sustainable manner (WWF 2007). At the same time, domestic policy interventions to promote and support the introduction of new technologies are required. We note the central importance of public policy in creating a demand for new technologies through the creation of markets and regulatory incentives, and in managing the risks associated with the adoption of these technologies through the provision of information; the development of codes, standards, and certification processes; and access to finance through risk guaranty mechanisms. Finally, we note that technology investment is not limited to physical capital investment, but also extends to personnel capacity-building, institution-building and infrastructure development.

Notes

1 For details, see http://www.ireda.in/default.asp.

2 Presently ranging from 20–60 watts, i.e. four to 10 per cent of total residential energy consumption.

3 In India, during 2005–06, the residential sector accounted for 13.6 per cent of final energy consumption (TERI 2007). This is exclusive of the energy used for transportation.

4 The Bureau of Energy Efficiency (BEE) is a statutory body under the Ministry of Power, Government of India. The mission of BEE is to institutionalize energy efficiency services, enable delivery mechanisms in the country and provide leadership to the key players involved in the energy conservation movement. The primary goal of the Bureau is to reduce the energy intensity in the economy. For details, see <http://www.bee-india.nic.in/aboutbee/aboutus.html>.

5 In the context of technology, this may imply the public purchasing of standard technology products, or 'developmental' procurement that calls for the creation of new products, or 'adaptive' public technology procurement, which requires existing technology to be adapted to specific local conditions (Edquist *et al.* 2000).

6 See Issues Paper prepared for a multi-stakeholder dialogue on 'Rural electricity access and distributed generation,' organized by TERI and S R Corporate Consultant, 9 January 2008, Raipur.

References

Chikkatur, A. P., Ambuj, D., Sagar, N. A., and Sreekumar, N., 2007. 'Tariff-based incentives for improving coal-power-plant efficiencies in India.' *Energy Policy* 35, pp. 3744–58.

Edquist, C., Hommen, L. and Tsipouri, L., 2000. 'Introduction' in Edquist, C., Hommen, L. and Tsipouri, L. (eds) *Public technology procurement and innovation.* Massachusetts: Kluwer.

EREC (European Renewable Energy Council) and Greenpreace International, 2007. *Future investment – a sustainable investment plan for the power sector to save the climate.* Viewed 17 May, 2008 <http://www.energyblueprint.info/fileadmin/media/documents/PressReleases/FutureInvestment.pdf?PHPSESSID=8382b5defe372c9e851aa9c4649b71b3>.

Jamasb, T., Nuttall, W. J. and Pollitt, M. G., 2006. 'New electricity technologies for a sustainable future.' In Jamasb, T., W J Nuttall, W. J. and Pollitt M. G. (eds) *Future electricity technologies and systems.* Cambridge: Cambridge University Press, pp. 1–23.

Mathur, A., 2007. 'Energy efficiency in buildings in India: an overview.' Presentation to the Second Meeting of the Indo–German Energy Forum, 20 December 2007, New Delhi.

McKinsey & Co, 2007. 'Curbing global energy demand growth: the energy productivity opportunity.' McKinsey Global Institute, May 2007.

Sathaye, J. and Murtishaw,., 2004. *Market failures, consumer preferences, and transaction costs in energy efficiency purchase decisions.* Consultant Report prepared for California Energy Commission, USA. Viewed 18 May, 2008 <http://www.energy.ca.gov/2005publications/CEC-500-2005-020/CEC-500-2005-020.PDF>.

Shahi, R. V., 2006. Keynote address at the India – International Energy Agency (IEA) Joint Workshop on Energy Efficiency in Buildings and Building Codes, 4–5 October 2006, New Delhi. Viewed 18 May, 2008 <http://powermin.nic.in/whats_new/pdf/ENERGY_EFFICIENCY_IN_BUILDINGS.pdf>.

TERI (The Energy and Resources Institute), 2007. *TERI energy data directory and yearbook.* New Delhi: TERI.

UNFCCC, 2008. 'CDM statistics,' website of the United Nations Framework Convention on Climate Change (UNFCCC). Viewed 19 May, 2008 <http://cdm.unfccc.int/Statistics/index.html>.

Vankani, P., 2005. 'Triggering energy efficiency markets: the role of ESCOs.' Presentation to the Second Task Force Workshop on Energy Conservation in the Petrochemical and Refinery Sectors, Indian Petrochemicals Corporation Limited (IPCL), 19–20 May 2008, Vadodara.

WWF, 2007. *WWF's vision for 2050, climate solutions.* WWF International.

Part IV

Nuclear energy for India – the debate

13 Nuclear power growth

An option for sustaining Indian energy requirements

Ravi B. Grover

Introduction

The world's population may be anywhere between nine and 10 billion by the middle of this century. Almost all the increase will take place in the developing countries. Access to energy is a prerequisite for growth. It increases productivity in agriculture and industry, as well as services, thereby acting as a multiplier for human effort. It contributes to social development by improving health and education. At present there is a large inequality in per capita energy consumption between and within nations, and energy supplies are under strain due to increasing demand and the depleting resources as economies grow. The criticality of energy supplies to the world can be gauged by the space devoted to issues related to energy in daily newspapers, visual media, and popular and academic journals. Energy-related issues are getting intertwined with politics and power[1] and history tells us (Podobnik 2006) they have always been so intertwined.

Energy will continue to dominate the world stage for all times to come. Issues related to energy are availability of fuel supplies; technology for their efficient utilization; and environmental impacts, particularly concerning climate change and health. No single energy resource or technology can be said to be a silver bullet. Each country or a group of countries has to plan and build an energy system best suited to its fuel resource profile, technology base and human resources. Sustainability of energy supplies and the environmental issues associated with sustainability demand that all non-carbon emitting resources become an integral part of an energy mix – as diversified as possible – to-ensure energy security to the world during the present century and beyond. In spite of the fact that the environmental implications of burning coal are well known, coal will continue to be the dominant fuel the world over for meeting growing energy demand. The only solution to mitigate environmental implications is to step up investment in research and development in carbon sequestration technologies and to deploy such technologies at the earliest. Investment in research and development is also needed for the increased exploitation of low-carbon fossil fuels, renewables and nuclear energy.

The average annual electricity consumption in the world is about 2,600 kWh per capita (International Energy Agency 2007) and the present population is about

Table 13.1 Actual power supply position in India

Year	*Peak demand (MW)*	*Peak met (MW)*	*Peak deficit (%)*	*Energy requirement (GWh)*	*Energy availability (GWh)*	*Energy deficit (%)*
2002–03	81492	71547	12.2	545983	497690	8.8
2003–04	84574	75066	11.2	559264	519398	7.1
2004–05	87906	77652	12.3	591373	548115	7.3
2005–06	93255	81792	12.3	631757	578819	8.4
2006–07	100715	87105	13.5	693057	624716	9.9

Source: Central Electricity Authority 2007a.

Note Energy availability figures do not include generation from captive power plants, which has been estimated as 78 TWh during 2006–07. Thus total generation during 2006–07 was above 700 TWh. (Ministry of Power 2007).

6.5 billion. Assuming that the population will stabilize at 10 billion and an average annual electricity consumption of about 5,000 kWh per capita will be sufficient to provide acceptable standards of living, globally electricity generation has to increase by three times and most of this increase must take place in the developing countries. India and China, the world's two most populous countries, are advancing fast economically and developments in the energy sector of these two countries will influence world energy markets and the global environment. India and China, due to their sheer size and consequent demand for energy, can also be expected to lead innovation in energy technologies (Podobnik 2006).

Focusing on India, the position of the electric power supply is very precarious throughout the country. Table 13.1 illustrates the demand and supply position. One may note that India has been facing energy shortages in the range of 7 to 10 per cent during the past five years. Several studies have forecast a robust growth in India's economy in the coming five decades, but it is necessary that infrastructure, and particularly energy infrastructure, is strengthened to support the anticipated economic growth. This has to be done in a manner sustainable from the standpoint of the availability of energy resources and the effect on the environment. This chapter briefly examines the current energy scenario, growth projections, energy supply options and the important role to be played by nuclear energy. Except for brief remarks on primary energy in the section on demand projections, this chapter deals with electricity.

Demand projections

To plan for growth in India's electricity requirements, the Central Electricity Authority in India has been periodically carrying out detailed surveys. These span about fifteen years each. The seventeenth survey was published in 2007 (Central Electricity Authority 2007). As per this survey, peak demand is expected to be 152,746 MW in 2011–12, 218,209 MW in 2016–17 and 298,253 MW in 2021–22. Corresponding installed generating capacity would be about 210,000

MW in 2011–12, 300,000 MW in 2016–17 and 410,000 MW in 2021–22. These figures may be compared with the peak met in 2006–07 (Table 13.1) to get an idea about the extent of generation capacity to be added. Because energy availability is important for economic growth and there is a large unmet demand for electricity, studies to develop long-term scenarios have been carried out in recent years. The author had the privilege of being associated with one exercise to develop an energy growth scenario spanning five decades (Grover and Chandra 2004). This was done with the specific objective of assessing the role to be played by nuclear energy in India. This has been followed by a comprehensive study by an expert committee on Integrated Energy Policy set up by the Government of India under the Planning

Table 13.2 Installed capacity in MW as on 31.03.2007

Hydro	34653.8
Thermal Coal	71121.3
Gas	13691.8
Diesel	1201.8
Nuclear	3900.0
Wind/renewable energy	7760.6
Grand total	132329.3

Source: Central Electricity Authority 2007a.

Notes
a This does not include benefits from projects in Bhutan.
b This also does not include captive power plants.

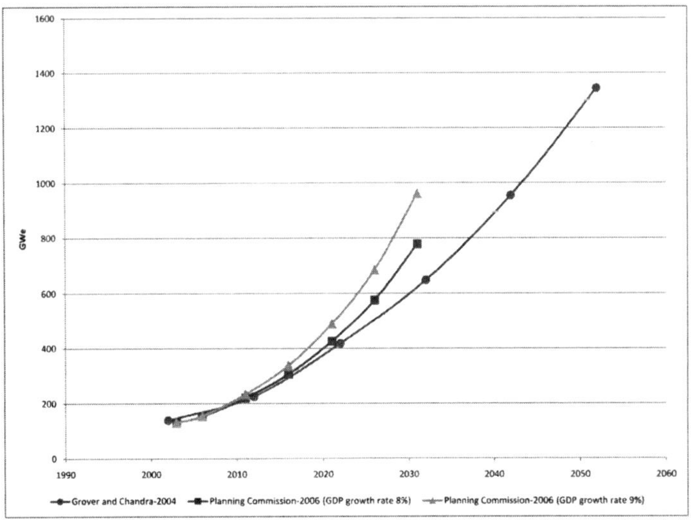

Figure 13.1 Scenarios for growth of India's installed electricity generating capacity.

Commission (Planning Commission 2006). Figure 13.1 gives the details of the scenarios developed by Grover and Chandra as well as the Expert Committee[2].

Grover and Chandra focused on the fuel-resource position and therefore assumed the peak to average power ratio as one. This ratio depends on several factors, particularly fuel mix and load variations. If one assumes this ratio[3] to be 1.25, installed capacity by the middle of the century could be as high as 1,700 GW. This represents a manifold increase over the installed capacity as on 31 March 2007 (Table 13.2). Grover and Chandra have estimated the generation in the year 2052–53 to be 8,000 TWh. This also is an order of magnitude higher than the generation of about 700 TWh in the year 2006–07. Such a large increase in electricity generation capacity calls for a serious examination of the fuel resource position, technologies that need to be developed and long-range planning. Based on various projections, Grover and Chandra expect the population of India in the year 2052–53 to be 1.5 billion. Generation as large as 8,000 TWh will be able to ensure a per capita availability of only about 5,000 kWh per annum. This is very modest as compared to the present average in the developed world.

One has also to look at growth projections from the standpoint of cumulative energy usage in the coming decades. Grover and Chandra estimate cumulative primary energy usage for five decades beginning from the year 2002–03 to be about 2,400 EJ. This needs to be compared with the energy resources available to India. The next section examines supply options.

Supply options

India is not well endowed with energy resources, at least not when compared to its requirements. Coal is the dominant resource and proven coal reserves, even if one assumes a modest growth in consumption at the rate of five per cent per year, will last only until the middle of this century. In 2006, China produced 2,481 tons, the United States 990 tons and India 427 tons (International Energy Agency 2007). While Indian coal production is much lower than that of China, its continuous increase would require investment in mining as well as railroad transportation. The continued depletion of coal reserves requires a focusing of attention on coal mining so that reserves considered uneconomic at present can be exploited. Oil and gas reserves are very modest and India is already importing most of its requirements. India is not a well explored country, however, and the recent stepping up of exploration for petroleum may yield positive results. From a long-term perspective, unconventional sources of gas such as coal-bed methane and gas hydrates could also be important for India (Sudarshan 2006).

Table 13.3 gives details with regard to hydro and other renewable resources. Hydro potential in India is reasonable and needs to be fully exploited. With regard to other renewable sources, the report of the Planning Commission says that the energy potential has been assessed independently for each option. If all such options are developed together, the combined potential may be less than the sum, due to a paucity of available land for energy generation as other, competing, land uses may dominate. The potential of solar as indicated in the table is significant,

Table 13.3 Renewable energy resources

Resources	*Unit*	*Present*	*Potential*
Hydro-power	EJ/year	1.019	4.734
Wood	EJ/year	5.862	25.958
Biogas	EJ/year	0.029	0.795
Biodiesel	EJ/year	–	0.837[a]
Bioethanol	EJ/year	< 0.04	0.419
Solar - Photovoltaic	EJ/year	–	50.241[b]
Solar - Thermal	EJ/year	–	50.241[c]
Wind energy	EJ/year	< 0.04	0.419
Small hydro	EJ/year	< 0.04	0.209

Source: Planning Commission 2005.

Notes
a Availability of land and inputs for getting projected yields is a critical constraint.
b Expected by utilizing 5 million hectares wasteland at an efficiency level of 15% for solar photovoltaic cells.
c MW scale power plants using 5 million hectares

Table 13.4 Population density of selected countries

Country	*Area in square km*	*Population in millions (2002)*	*Population density in persons/sq km*
Canada	9,976,140	31.902	3.20
Argentina	2,766,890	37.813	13.67
Brazil	8,511,965	176.03	20.68
USA	9,629,091	280.562	29.14
EU-25	3,976,342	454.48	114.30
China	9,596,960	1,284.30	133.82
India	3,287,590	1,045.85	318.12
Japan	377,835	126.975	336.06

but on the basis of currently known technologies it is not cost competitive and that is why hardly any potential has been realized. Biodiesel and bioethanol derived from energy crops are being harnessed to a significant extent in some countries, but growing energy crops requires not only land, but also water and sunshine. These options have significant potential in countries that are not densely populated, such as Argentina and Brazil, but not in a densely populated country like India. Table 13.4 gives the population density of selected countries and the numbers tell the story. In a country like India there is going to be direct competition between energy crops and food crops – and food crops will always get higher priority. However, India must exploit the full potential of energy crops, however limited in percentage terms this share may be.

Considering India's existing installed capacity base, and the constraints to enhancing the share of other energy resources, *it is obvious that fossil fuels, particularly coal, will continue to be a dominant energy source in India during the near term.* Intensifying exploration to locate additional resources, particularly of petroleum, deploying modern technologies to improve recovery of known resources, and developing techniques like in situ coal gasification to exploit deep-seated coal reserves can all increase the net availability of fuel resources. Further, the most efficient generation technologies have to be deployed to make the best use of all available fuel resources.

Based on currently known technologies, nuclear is the only other supply option available. The fact that India is not well endowed with energy resources has been known ever since independence (1947) and the country has long invested in developing the infrastructure for harnessing nuclear technology. Yet, here again, India has a resource constraint in terms of her modest domestic reserves of uranium and constraints on international trade in uranium arising from the prevailing control regime.

For this reason India has planned to develop a closed fuel cycle, which makes use of the full potential of given nuclear fuel resources and also makes it possible to use thorium, which India has in abundance. Table 13.5 gives details about nuclear fuel resources and their energy potential when used in a closed fuel cycle and in an open fuel cycle.

In regard to nuclear resources and plans for their use, India has formulated a three-stage nuclear power programme involving the setting up of pressurized heavy water reactors (PHWR) in the first stage, fast breeder reactors in the second stage and reactors based on thorium fuel in the third stage. During each stage, associated fuel cycle facilities are also to be set up and the three stages are linked through such facilities. The three stages have to be followed sequentially, following a plan decided by fission parameters of the fuels and optimized to meet India's energy requirements.

The extent of breeding in a reactor depends on the type of reactor system i.e. thermal or fast; element used as fuel i.e., uranium or thorium and the chemical form of the fuel i.e., metallic, carbide, nitride or oxide. For a comparison between different reactor systems, the term 'breeding ratio' (Glasstone and Sesonske 1967) is used. It is defined as the ratio of the number of fissile atoms produced to the number of fissile atoms that have been consumed. The fission parameters

Table 13.5 Nuclear fuel resource position and its energy potential

Uranium metal available for fuel fabrication	*61000 tons*
Energy potential in Pressurised heavy water Reactors	2875 TWh
Energy potential in fast breeder reactors	370000 TWh
Thorium metal available for fuel fabrication	*225000 tons*
Energy potential when used in breeders	1363000 TWh

Source: Grover and Chandra 2004.

Table 13.6 Near-term (until 2020) nuclear power programme

Scenario I

Reactor type and capacities	*Capacity (MW)*	*Cumulative capacity (MW)*
17 reactors at 6 sites under operation at Tarapur, Rawatbhata, Kalpakkam, Narora, Kakrapar and Kaiga	4,120	4,120
3 PHWRs under construction at Kaiga (1×220 MW), Rawatbhata (2×220 MW)	660	4,780
2 LWRs under construction at Kudankulam (2×1000 MW)	2,000	6,780
PFBR under construction at Kalpakkam	500	7,280
Projects planned till 2020 – PHWRs (8×700 MW), FBRs (4×500 MW), AHWR (1×300 MW)	7,900	15,180

Scenario II

Scenario I as above plus additional import of 6 light water reactors, 1000 MW each, leading to a cumulative installed capacity of 21,180 MW by the year 2020.

of thorium do not allow for a high enough breeding ratio and therefore the introduction of thorium has to wait until a large enough installed capacity base has been established, based on uranium-plutonium fuelled reactor systems.

Table 13.6 shows the near-term nuclear programme in India. It provides two scenarios. While scenario I is based on indigenous resources and imports already tied up, realization of scenario II will depend on opening up of international civil nuclear cooperation and trade with India. Scenario I envisages the setting up of eight additional PHWRs of 700 MW rating, and four additional fast reactors of 500 MW rating. The Nuclear Power Corporation of India Limited is ready with the design of 700 MW PHWRs and can start construction as soon as more uranium mines have been opened up. The Department of Atomic Energy is making efforts to open new uranium mines and also to intensify uranium exploration in India. It should be possible to plan for more PHWRs if exploration yields positive results. The near-term plan also envisages the setting up of additional fast reactors after the start of operation, in the year 2011, of the prototype fast breeder reactor (PFBR), now under construction.

Scenario II envisages the import of six reactors in addition to the ongoing indigenous programme. It was drawn up in the year 1995 and because of electricity shortages it is desirable to have plans for higher additional capacity. Scenario II,

Table 13.7 Nuclear installed capacity growth possibilities until 2052 based on fast reactors (corresponding to scenario II of table 13.6)

Fuel type	*Out of pile period (years)*	*System doubling time (years)*	*Fast reactor installed capacity (MW)*	*Total installed capacity (MW)*	*Total nuclear generation (TWh)*
Oxide	Two	25.8	55	67	498
Oxide	One	18.9	81	93	691
Carbide	Two	14.7	90	102	758
Carbide	One	11.0	183	195	1,449
Metal	Two	12.3	115	127	944
Metal	One	8.9	263	275	2,044

Source: Chandra et al. 2007.

or improvements on those projections for nuclear energy, is possible only if the international nuclear trade opens up. The dialogue on this issue continues between India and other countries.

Reprocessing of spent fuel from PHWRs in reprocessing plants gives plutonium-239, which is the fuel for fast reactors. The rate of addition of fast-reactor installed capacity depends on the build-up of the plutonium inventory, which in turn depends on the choice of fast reactor fuel cycle. The PFBR under construction uses oxide fuel. Research and development has been launched to complete the development of metallic fuel by the year 2020. The fast reactors to be set up after the year 2020 will be based on metallic fuel.

Table 13.7[4] provides the results of a parametric study to determine nuclear installed capacity growth possibilities until the middle of this century. For minimum system doubling time, it becomes necessary to use metallic fuel and ensure that out-of-pile period is as low as practically possible. The maximum fast reactor capacity actually realized will depend on the success of the development of technologies necessary for deployment of metallic fuel and associated fuel cycle facilities, and the setting up of necessary infrastructure, particularly fuel reprocessing and fuel fabrication facilities.

The possible generation from nuclear fuel could be as much as a quarter of the total annual generation as per the scenario of Grover and Chandra (2004). *Thus, seen from a medium-term perspective, nuclear could play a significant role if fast reactors having desired characteristics are developed.*

Economics

Considering India's fuel resource endowments, population density and the need to provide energy for economic growth, a growth in the share of nuclear energy in India's energy mix is inevitable and also desirable due to global warming considerations. Nuclear power generation does not directly involve carbon dioxide emissions or emissions of other greenhouse gases. As India proceeds with the development of nuclear energy, its economic competitiveness also needs to be examined. There is a large body of literature on this topic and views are quite divergent.

A comprehensive study was published in 2005 by the Nuclear Energy Agency and the International Energy Agency, with the objective of providing reliable information on key factors affecting the economics of electricity generation using a range of technologies including nuclear (Nuclear Energy Agency and International Energy Agency 2005). There have been several national studies, some reviewed in Proust (2005). Proust examines the DIDEME report from France, MIT and University of Chicago studies from the United States, the TARAJANNE study from Finland and other studies. He concludes, 'All recent European cost studies show that third generation nuclear is competitive with coal-fired plants, and may be up to 20 per cent cheaper than CCGTs[5] for base load electricity generation, even when CO_2 emissions costs are disregarded.' He continues, 'This EU picture should also apply to the United States once the first new nuclear plants will have been successfully built and operated in the country.'

The situation in India is even more comforting than in the EU because the Nuclear Power Corporation of India Limited has been able to reduce the capital cost and gestation period of nuclear power plants (Thakur and Chaurasia 2005)[6]. Furthermore, while the cost of nuclear power is location independent, this is not so in the case of coal-based plants, where costs are very sensitive to the distance of the power plant from the coal mine. Therefore, nuclear power plants are being set up in India at locations that are distant from coal mines.

Since India proposes to set up several fast reactors in the years to come, it is necessary to examine the economics of fast reactors. Here again views are divergent with regard to the need for fast reactors, their economics and concerns about proliferation. Recent opinions are moving in favour of fast reactors. As described in the previous section, India's fuel resource position calls for the early introduction of fast reactors. A study carried out by the French utility, EdF, envisages the industrial deployment of a first series of fast reactors by around 2040 (Carre 2005). A Russian viewpoint (Anonymous 2005) is that 'mass-scale construction of fast reactors shall not be delayed any longer' as reserves of both 'cheap and costly uranium will be exhausted between 2030 and 2050'. They advocate finishing development work for the next generation of fast reactors within a decade and starting batch production of fast reactors by 2020. Another study from France by David (2005) goes a step closer to the Indian viewpoint and advocates development of thorium systems using fast breeder reactors and molten salt reactors to extend the energy potential of known reserves of nuclear fuel.

The divergence in views on the economics of fast reactors arises from the fact that very few countries have experience in operating commercial-scale reprocessing plants, and a repository for disposal of high-level waste is yet to be built. A study carried out by Bhoje (2003) with regards to the situation in India concludes that the cost of fast reactors will be comparable to, if not less than, PHWR cost. Bunn *et al.* (2003) examined these issues and concluded, 'Reprocessing and recycling plutonium in fast reactors (FRs) with an additional capital cost of \$200/kW (electric) compared to new LWRs will not be economically competitive with a once-through cycle in LWRs until the price of uranium reaches some \$340/kg U, given our central estimates of the other parameters. Even if the capital cost of new FRs could be reduced to equal that of new LWRs recycling in FRs would not be economic until the uranium price reached \$140/kg U.'

The price of uranium is already above \$140/kg U and Indian studies indicate that the capital cost of fast reactors will be comparable to PHWRs. A study by the Massachusetts Institute of Technology (Deutch and Moniz 2003) points out that reprocessing and recycling would impose a significant penalty on nuclear power. This conclusion has been contested by the French, who have significant industrial experience in reprocessing. Studies by the French Commissariat à l'Energie Atomique (CEA) (Arthur de Montalembert, 2003, personal communication) say that '... the incremental cost of MOX recycling is between four per cent and six per cent of the kWh cost.' One can extend this observation to fast reactor economics as well.

Studies published in the seventies, such as the Report of the Nuclear Energy Policy Study Group (1977), were categorical in saying that there is no case for reprocessing and recycling. Some even now persist with similar arguments (Suchitra and Ramana 2006) on considerations of economics and the environment. The security risks associated with having large stockpiles of separated plutonium are also cited as a reason against closing the fuel cycle. While the economics of reprocessing has been proved beyond doubt by the French experience, the increase in uranium prices since the turn of the present century, and the continued demand for more energy sources, have made it important to exploit the full energy potential of uranium by reprocessing spent fuel and recycling recovered plutonium. An economic argument following the principle of 'reprocess to reuse' and therefore, reprocessing need not be associated with stockpiles of plutonium. A report by the Nuclear Energy Study Group of the American Physical Society Panel on Public Affairs says:

> In the longer term, the balance among the benefits, costs, and risks of reprocessing may change significantly. By reprocessing spent fuel and burning the recovered uranium and plutonium in a nuclear "breeder" reactor, it is possible to get as much as 50 times more energy out of the original uranium. Therefore, if nuclear energy expands substantially in the future and puts pressure on the availability of low-cost uranium fuel, then reprocessing and breeder reactors could become the preferred option if the associated proliferation risks can be addressed.
>
> (American Physical Society 2005)

Direct disposal of spent fuel implies creating a plutonium mine, with unknown future consequences, and in this author's opinion, is an irresponsible option. Reprocessing and recycling does not leave such unresolved issues for the future and is an option that should be pursued by all nuclear-capable nations so as to preclude unknown consequences as well as to ensure the sustainability of energy resources.

The pursuit of a closed fuel cycle approach makes nuclear power a sustainable technology with known fuel resources that will be sufficient for several centuries. Its safety record is superior compared to other technologies, having capacity for expanded supplies. It emits no carbon dioxide or other greenhouse gases. Its costs are competitive and it has to be an essential part of the energy mix of a large country like India.

Safety

Safety is a topic that engages the attention of everyone. The nuclear industry all over the world has been conscious of safety and has been making efforts to enhance safety. If we compare various methods of electricity generation, the nuclear industry has the best record. Immediate fatalities for the period 1970–92, normalized to deaths per TWy of electricity generated, are 342 for coal, 85 for natural gas, 883 for hydro and eight for nuclear (Anonymous 2005a). The nuclear industry is the only industry where operators from all over the world have come together and set up an institution, the World Association of Nuclear Operators (WANO). This association conducts peer reviews of all nuclear power plants and is evolving industry-wide procedures and practices to continuously enhance safety.

From considerations of economics, sustainability and energy security, a large-scale augmentation of nuclear generating capacity in India is desirable. It is a safe technology, as demonstrated by about 13,000 reactor-years of cumulative commercial operation. As per the IAEA database, at present 438 reactors are operating in 30 countries and another 31 reactors are under construction. There have been only two major accidents in the history of nuclear power. The first occurred at Three Mile Island, in the United States, in 1979. The reactor was damaged, but there were no adverse health effects or environmental consequences. The second occurred at Chernobyl, Ukraine in 1986, where the destruction of the reactor by explosion and fire killed 31 people immediately. The report by the Chernobyl Forum (IAEA 2005) says that as of mid-2005, fewer than 50 deaths had been directly attributed to radiation from the disaster. Burton Bennett, Chairman of the Chernobyl Forum explains:

> This was a very serious accident with major health consequences, especially for thousands of workers exposed in the early days who received very high doses, and for the thousands more stricken with thyroid cancer. By and large, however, we have not found profound negative health impacts to the rest of the population in surrounding areas, nor have we found widespread contamination that would continue to pose a substantial threat to human health, with a few exceptional, restricted areas.

Concluding remarks

Providing electricity to fuel India's economic growth is a challenge for planners, given the constraints on fuel resource availability. As coal and hydrocarbons are depleted, it will become necessary to develop nuclear technology and improve the share of nuclear electricity in the years to come. Considering the size of India's population, conclusions with regard to India have global implications.

The Economist in its 8–14 September 2007 issue says, 'Geopolitics, technology, economics and the environment are all changing in nuclear power's favour.' It continues, 'Simpler designs cut maintenance and repair costs. Shut-downs are now far less frequent, so that a typical station in America is now online 90 per cent of the time, up from less than 50 per cent in the 1970s.' Most significantly, green gurus such as James Lovelock, Stewart Brand and Patrick Moore have changed their minds and embraced nuclear power. New builds are already competitive in several countries and this author believes that the economics of nuclear power in countries where the nuclear industry has been moribund will change once the first few plants have been built.

Notes

1 The special edition of Newsweek: December 2006–February 2007–Issues 2007 focuses on energy and takes the reader 'on a tour of the startling changes underway, including the new geopolitics of oil, and how the imminent rise of natural gas will redraw the balance of power once again.' It carries a two-page spread with a caption, 'Politics, Power and Petroleum' and this caption sums up the present energy scenario.

2 Grover and Chandra concentrated on energy resource usage and therefore considered the ratio of installed capacity required to projected peak demand as one. Appropriate factors have been considered in the study by the Planning Commission.

3 In the estimates given in the report of the expert committee (Planning Commission 2006), the ratio of installed capacity to peak demand reduces from 1.47 in 2003–04 to 1.31 in 2031–32.

4 Two new terms viz., 'system doubling time' and 'out of pile' period used in this table need to be explained. In its simplest form, the doubling time is defined as the time required for a breeder reactor to produce a surplus amount of fissile material equal to that required for the initial charge of the reactor (Glasstone and Sesonske 1967, 703). This definition does not take into consideration the time the material spends in other parts of the fuel cycle, e.g., transportation, intermediate storage, pretreatment and reprocessing of spent fuel and refabrication of plutonium into fuel. The time spent outside the reactor is called 'out of pile time'. 'The system doubling time' is that in which surplus fissile material produced would equal the total quantity in the fuel cycle (reactor + out of pile).

5 Combined Cycle Gas Turbine.

6 Economics of nuclear power from heavy water reactors have been disputed by Ramana *et al.* (2005) and Ramana (2007). They have drawn their conclusions by comparing the levelised cost of generation from nuclear plants, Kaiga 3 and 4, with the corresponding cost of generation from coal-fired power plants at Raichur (unit 7). The analysis by Ramana *et al.* is based on specific plants and is not a generic study. The basic data used have also been disputed by the Nuclear Power Corporation of India Limited (Thakur, 2005).

References

Anonymous, 2005. *Russian & NIS nuclear fuel cycle front-end industry*. Moscow: International Business Relations Corporation, July – September.

Anonymous, 2005a. 'Safety of nuclear power reactors.' Nuclear Issues Briefing Paper

14. Melbourne, Australia: Uranium Information Centre Ltd.
American Physical Society, 2005. 'Nuclear power and proliferation resistance: securing benefits, limiting risk.' Report by the Nuclear Energy Study Group of the American Physical Society Panel on Public Affairs.
Bhoje, S. B., 2003. 'Status of fast reactor development in India.' International Conference on Nuclear Power Technologies with Fast Neutron Reactors, Obninsk, 8–12 December 2003.
Bunn, M., Holdren, J. P., Fetter, S. and Zwaan, B V D., 2005. 'The economics of reprocessing versus direct disposal of spent nuclear fuel.' *Nuclear Technology*, 150, 209–230.
Carre, F., 2005. 'Fast reactors R&D strategy in France for a sustainable energy supply and reduction of environmental burdens.' *JAIF International Symposium*, Tokyo, 24 March 2005.
Central Electricity Authority, 2007. *17th electric power survey of India*. New Delhi: Central Electricity Authority, Government of India.
Central Electricity Authority, 2007a. *Power scenario at a glance*. New Delhi: Central Electricity Authority, Government of India.
Chandra, S., Srivastava, A. and Grover, R. B., 2007. 'Energy security through nuclear energy in India.' *Nuclear India*, 40(7–8), pp. 3–11.
David, S., 2005, 'Future scenarios for fission based reactors.' *Nuclear Physics A* 751, 429c–441c.
Deutch, J. M. and Moniz, E. J., 2003. 'The future of nuclear power: an interdisciplinary MIT study.' Massachusetts Institute of Technology, July.
Grover, R. B., and Chandra, S., 2004. 'A strategy for growth of electrical energy in India.' Mumbai: Department of Atomic Energy document no. 10. Also published as 'Scenario for growth of electricity in India', *Energy Policy*, 34 (2006), pp. 2834–47.
Glasstone, S., and Sesonske, A., 1967. *Nuclear Reactor Engineering*. New York: Van Nostrand Reinhold Company, p. 703.
International Energy Agency, 2007. *Key World Energy Statistics*.
IAEA, 2005. Press release by IAEA, WHO and UNDP on the release of the report, 'Chernobyl: the true scale of the accident; 20 years later: a UN report provides definitive answers and ways to repair lives.' 5 September.
Ministry of Power, 2007. *Report of the working group on power for the Eleventh Plan (2007–12)*. New Delhi: Ministry of Power, Government of India.
Nuclear Energy Study Policy Group, 1977. 'Nuclear power issues and choices.' Sponsored by Ford Foundation and administered by Mitre Foundation, Ballinger Publishing Company.
Nuclear Energy Agency and International Energy Agency, 2005. *Projected costs of generating electricity: 2005 update*. Paris: International Energy Agency, OECD.
Planning Commission, 2006. *Integrated energy policy: report of the Expert Committee*. New Delhi: Planning Commission, Government of India.
Podobnik, B., 2006. *Global energy shifts*. Philadelphia: Temple University Press & TERI Press, 9–11; 169.
Proust, E., 2005. 'Economic competitiveness of new (3rd Generation) nuclear plants: a French and European perspective.' *Proceedings of ICAPP05*, Seoul, Korea, 15–19 May 2005.
Ramana, M. V., D'Sa, A. and Reddy, A. K. N., 2005. 'Economics of nuclear power from heavy water reactors.' *Economic and Political Weekly*, 23 April, 1763–73.
Ramana, M. V., 2007. 'Economics of nuclear power: subsidies and competitiveness.' *Economic and Political Weekly*, 13 Jan, 169–171.
Suchitra, J. Y. and Ramana, M. V., 2006. 'High costs, questionable benefits of reprocessing.' *Economic and Political Weekly*, 25 November, 4848–51.
Sudarshan, A., 2006., 'Unconventional sources of gas: a short review.' *Energy Security Insights*, TERI, 1(2), pp. 17–26.
Thakur, S., 2005. 'Economics of nuclear power in India.' *Economic and Political Weekly*, 3 December, 5209.

Thakur, S. and Chaurasia, B. P., 2005. 'Cost effectiveness of electricity generating technologies.' NPCIL report, September, 2005. Nuclear Power Corporation of India Limited.

14 The many phases of nuclear insecurity

M. V. Ramana and J. Y. Suchitra

Over the last couple of decades, most discussions on energy policy have tended to advance energy security as an important consideration. In India, the Planning Commission's Integrated Energy Policy defines energy security as the ability to 'supply lifeline energy to all our citizens irrespective of their ability to pay for it as well as meet their effective demand for safe and convenient energy to satisfy their various needs at competitive prices, at all times and with a prescribed confidence level considering shocks and disruptions that can be reasonably expected' (Planning Commission 2006). Based on this definition, it argues in favour of rapid expansions in all forms of electricity generation, including nuclear power.

In the context of nuclear power, there are at least two problems with the definition as well as with the way it has been interpreted in the energy plan that the commission has developed. First, economic viability seems to be the only consideration for an energy-secure country. There is no mention whatsoever in the definition about environmental and other consequences of such energy production and the need to ensure that people are not harmed directly or indirectly through the process of energy production.[1] Thus, although equity considerations are included in the supply of energy, the process of production results in inequitable impacts.

A second problem is that when potential energy sources are listed, no economic considerations are offered – thereby negating the criterion of competitive prices in its own definition. For example, with reference to nuclear energy, it argues that 'India has to succeed in realizing the three-stage development process … and thereby tap its vast thorium resource to become truly energy independent beyond 2050. Continuing support to the three-stage development of India's nuclear potential is essential'.[2] As we shall show, even the breeder reactors that constitute the second stage are going to be an expensive source of electricity and certainly not cost-competitive. The third stage will only be more uneconomical. Likewise, while the requirement posited is for 'safe energy', there is no consideration of the known safety problems with nuclear reactors. Among all electricity-generating technologies, nuclear energy alone is prone to creating catastrophic accidents with a potentially global impact.

Nuclear power, which official energy planners have encouraged, is also not environmentally sustainable. It is a highly polluting source of power due to the radioactivity released into the environment at almost every stage of the nuclear fuel cycle. Some radioactivity will continue to pose risks to human health for hundreds of thousands of years because there is as yet no satisfactory solution to the radioactive waste problem.

Finally, another unique aspect is the close connection between nuclear energy and nuclear weapons. Despite the development of elaborate safeguarding schemes, there remains the strong possibility of using the infrastructure and knowledge developed to generate nuclear electricity to manufacture nuclear weapons. Thus, nuclear power is also linked to military insecurity.[3]

In this chapter, we focus primarily on the high economic costs of nuclear power in India, along with a brief overview of its environmental and safety implications. These arguments show that nuclear power cannot be a candidate for ensuring energy security.

Nuclear economics in India

Plans by the Department of Atomic Energy (DAE) to expand nuclear power in India revolve around a three-phase nuclear programme envisaged in the 1950s (Bhabha and Prasad 1958). The first phase involves the use of uranium fuel in pressurized heavy water reactors (PHWR), the second phase involves fast breeder reactors that use plutonium from reprocessed spent fuel from PHWRs and uranium and thorium to produce plutonium and uranium-233, and the third phase involves breeder reactors using uranium-233 and thorium. Though it has been over five decades since the three-phase programme was announced, all of the DAE's operating commercial reactors fall under the first phase. It is only just starting on phase two and is constructing the country's first industrial-scale breeder reactor. In the next few sections, we look at the economics of the first two phases, and that of reprocessing, a necessary component in this strategy.

Table 14.1 Cost and other figures for heavy water reactors and thermal plant

	Kaiga I & II	*Kaiga III & IV*	*RTPS VII (coal)*
Power plant capacity (MWe)	440	440	210
Total construction cost[1] (billion mixed year Rs)	18.16	27.27	4.91
Overnight construction cost (Rs/kWe)	75,150	53,969	26,914
Present value at 5% discount rate (billion Rs)	45.08	31.20	6.09
Uranium (mining + fabrication) price (Rs/kg)	17,934	17,934	
Heavy water price (Rs/kg)	27,124	27,124	
Decommissioning cost (fraction of capital cost)	10%	10%	
O & M cost (fraction of capital cost)	2%	2%	2.5%
Coal cost (Rs/ton)			1,539

Note: All figures in 2004 Rupees unless noted otherwise.

1 This is the sum of the actual expenditures incurred (or projected to be incurred, in the case of the Kaiga III & IV) in mixed year figures and does not include interest during construction (IDC).

Economics of pressurized heavy water reactors

Since the 1950s, the DAE has claimed that the cost of nuclear power from PHWRs compares very favourably with electricity from coal-fired thermal power plants, India's staple source of electricity (Bhabha and Prasad 1958; Srinivasan 1985; Nema 1999; Thakur and Chaurasia 2005). However, these claims are based on projected figures rather than the actual costs of constructed nuclear and thermal plants. In contrast, an analysis of the cases of Units I and II and Units III and IV of the Kaiga Atomic Power Station and Unit VII of the Raichur Thermal Power Station, using the standard discounted cash-flow (DCF) methodology,[4] shows that for a wide range of realistic parameters, nuclear power is significantly more expensive (Ramana, D'Sa, and Reddy 2005; Ramana 2007). The major cost components and other figures are listed in Table 14.1. The coal is assumed to come from mines that are 1,400 km away.

A particularly important variable is the discount rate, which is a measure of the value of capital. Nuclear power, being a very capital-intensive technology, is competitive only for low discount rates (see Figure 14.1). But given multiple demands on capital for infrastructural projects, including for electricity generation, very low discount rates are not realistic.[5] At a real discount rate of five per cent, roughly what is recommended by the Central Electricity Regulatory Commission (CERC) (CERC 2006), nuclear power from the Kaiga III and IV reactors is about eight per cent more expensive than thermal power from RTPS VII.

The RTPS VII case is somewhat atypical in that it includes a large transport cost for coal because of the assumption that the coal is sourced from mines that are 1,400 km away. Over a third of all of India's coal plants are at the pithead and a further one quarter or more are within 500 km of one (Chowdhary 1998). In sum, nuclear power will generally be far more expensive than thermal power. A larger proportion of nuclear capacity therefore implies that the overall supply of electricity becomes more expensive. Poorer sections of society will not be able to afford electricity, at least without greater subsidies, which would be detrimental to energy security.

This economic comparison is largely based on assumptions favourable to nuclear power. For example, the comparison does not include any insurance liability against accidents since the government has not required that of nuclear power plants. Most important, following the methodology adopted by the DAE (Thakur and Chaurasia 2005), we have not included the costs of dealing with radioactive wastes from nuclear power. In essence, the Nuclear Power Corporation (NPC), which operates the heavy water reactors, simply hands over the irradiated spent fuel from its rectors to the DAE. However, since reprocessing is a service rendered by the DAE to the NPC, the rational choice for the DAE would be to charge a fee for the same. By not doing so, the DAE, in effect the taxpayer, is offering the Nuclear Power Corporation a subsidy.[6]

The DAE treats spent nuclear fuel by reprocessing it and segregating the waste into different categories on the basis of their radioactivity. As we discuss in the following section, reprocessing is expensive. If even half of our estimate of the

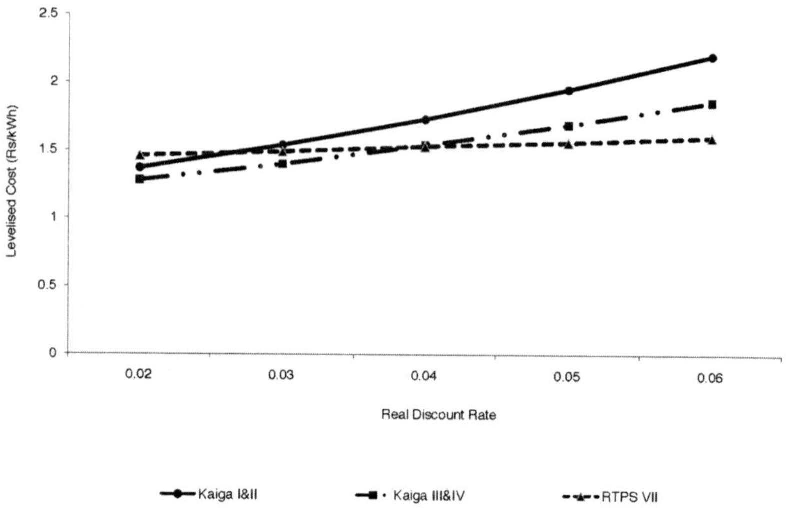

Figure 14.1 Busbar generation costs of Kaiga I and II, Kaiga III and IV (projected costs), and RTPS VII at 80 per cent capacity factor.

cost of reprocessing in India is included in the tariff for nuclear power, it would make it 25 per cent more expensive than thermal power from coal.

The Indian case is by no means unique. In the country with the most nuclear reactors, the United States, a detailed study conducted at the Massachusetts Institute of Technology found that unless there are dramatic improvements in nuclear cost factors (and none in the other technologies), nuclear power is simply not competitive with the other technologies (Deutch *et al.* 2003). Even with relatively optimistic assumptions about construction costs and times for nuclear plants, the MIT study's model estimated a real levelized cost of 6.7 cents/kWh for nuclear power, 4.2 cents/kWh for coal, and 3.8–5.6 cents/kWh for gas. Investors would naturally shy away from nuclear power.

Economics of reprocessing

As mentioned earlier, the economics of nuclear power is significantly affected by the cost of dealing with the irradiated spent nuclear fuel. Spent fuel can either be reprocessed or disposed of directly. Direct disposal involves long-term storage of the spent fuel followed by its encapsulation and permanent storage in a geological repository. No country in the world has yet built a geological repository.

Reprocessing involves the chemical processing of the spent fuel to separate out the plutonium and the (depleted) uranium contained. The plutonium thus extracted is used to fabricate fuel for nuclear reactors.[7] Reprocessing, therefore, provides both a service – that of dealing with the spent fuel, as well as a product – plutonium to fuel reactors. Reprocessing also produces high-level radioactive waste, which is vitrified and put into long-term storage. As with direct disposal, the plan is to

Table 14.2 Cost components and other assumptions

KARP construction cost (mixed year Rs)	5.58 billions
Overnight construction cost	12.86 billions
Present value at 5% discount rate	25.78 billions
Annual O&M Expenses	94 millions
Waste immobilisation plant construction cost (mixed year Rs)	499.9 millions
Overnight construction cost	912 millions
Present value at 5% discount rate	1.64 billions
Annual O&M expenses of WIP	62 millions
Storage facility (S3F) construction cost (mixed year Rs)	173.8 millions
Overnight construction cost	664 millions
Present value at 5% discount rate	1.35 billions
Transportation of vitrified waste	0.135 millions/ton
Geological disposal of vitrified waste	0.2 millions/ton
Decommissioning of KARP	4.12 billions
Interim storage of spent fuel before direct disposal	0.37 millions/ton
Transportation of spent fuel	0.3 millions/ton
Geological disposal of spent fuel	0.45 millions/ton

Note: All figures in 2004 Rupees unless noted otherwise.

bury the waste in geological repositories. There are also intermediate and low level radioactive wastes that are disposed off in other ways into the environment.

For decades now, there has been a debate on which of these options is cheaper. The general consensus based on cost data from Western countries is that reprocessing as a waste management technique is far more expensive than direct disposal (National Research Council 1996; Bunn *et al.* 2005; Deutch *et al.* 2003). This is primarily because of the enormous capital cost of the reprocessing facility.

Although no rigorous study has been conducted to support its assertions, the DAE claims that it is an exception to international cost norms (Balu and Ramanujam 1999; Sethna 1979). Indeed, it has maintained that '*economic considerations dictated the need for spent fuel reprocessing in India*' (Prasad 1996) (our emphasis). We examined the validity of this statement by looking at the cost of reprocessing at the Kalpakkam Atomic Reprocessing Plant (KARP) facility (Ramana and Suchitra 2007). KARP is used as the reference because it is the most recently constructed plant and is to serve as a standard design for future plants (Dey 2003).

Using the DCF approach, we find that reprocessing is not cost competitive with direct disposal as a waste management mechanism (Ramana and Suchitra 2007). The data that have gone into our calculation of the cost of reprocessing and direct disposal are summarized in Table 14.2.

Based on these figures, at a real discount rate of 5 per cent, our estimate of the total cost of reprocessing is Rs. 25,983 per kg of spent fuel (SF). Assuming that 99 per cent of the plutonium is recovered, the cost of producing each gram of plutonium is Rs. 6,999. The cost of reprocessing depends sensitively on the efficiency with which the plant operates. The above figures assume the highly optimistic value of 80 per cent.[8] If the capacity factor were to even come to 70 per cent, the reprocessing cost would go up to Rs. 29,569/kgSF.

Our calculations also show that direct disposal will cost only Rs. 1,120/kgSF. Clearly, as a waste management option, reprocessing is much more expensive than direct disposal. All of this nullifies the DAE's claims that reprocessing is dictated by economic considerations.

What could make reprocessing economically viable is if the plutonium were to be used in reactors to produce electricity, and that turns out to be cheaper than producing electricity from reactors using uranium fuel. It has been established that this is not the case with Western reactors and associated facilities and will not be until uranium prices are much higher than current values[9] (Bunn *et al.* 2005). We next test this proposition for the prototype fast breeder reactor (PFBR) that is currently being constructed in India.

Economics of prototype fast breeder reactors

While many countries were initially enthusiastic about breeder reactors, most have given up on them (Von Hippel and Jones 1997). Nevertheless, the DAE has been pursuing this sort of reactor without ever re-examining the feasibility of its initial vision. The argument offered for this pursuit is that India has only 'modest uranium reserves' of about 60,000 tons, 'which can support 10,000 MWe (megawatt electric) of PHWR capacities' (Kakodkar 2006). While widely articulated, this formulation is misleading. India's uranium resource base cannot be represented by a single number. As with any other mineral, at higher prices it becomes economic to mine lower grade and less accessible ores. Exploiting these would increase the amount of uranium available. Therefore, uranium resources can only be specified as a function of price.

As a way of evaluating the economics of breeder reactors, we compare the cost of generating electricity at the PFBR with a PHWR, the mainstay technology of the Indian nuclear programme. In order to address the argument about India's limited uranium reserves we undertake this as a function of uranium price and calculate the crossover price when the two technologies generate electricity at the same cost.

The PHWRs that we choose are Kaiga III and IV, whose cost figures are mentioned earlier. We consider the same set of cost components for both PFBRs and PHWRs, namely capital cost of constructing the reactor, fueling costs, operations and maintenance, decommissioning, and the management of low level wastes. In line with the DAE's methodology, we do not include the cost of dealing with the spent fuel; however, in the case of the PFBR, because its entire rationale is to generate more plutonium than it consumes, the plutonium used to fuel the reactor has to be derived from reprocessing its own spent fuel. The exception to the latter is for the initial loading of plutonium and the first few reloads; these have to come from plutonium recovered from reprocessing PHWR spent fuel in KARP

Table 14.3 Cost of electricity from breeder and heavy water reactors

	PFBR	*Kaiga III & IV*
Sum of annual construction costs (Billion Mixed year Rs)	34.92	27.27
Overnight construction cost (Rs/kW)	56,164	60,335
Present value of capital cost (Billion Rs)	22.80	25.65
Decommissioning cost (fraction of capital cost)	10%	10%
O & M cost (fraction of capital cost)	2%	2%
Capacity factor	80%	80%
Present value of lifetime fuel cost (Billion Rs)	64.74	21.28
Total Lifecycle cost (Billion Rs)	99.73	67.68
Levelised Cost (Rs/kWh)	2.37	1.66
Levelised Cost (cents/kWh)	5.40	3.78
Percentage Difference (PFBR-PHWR)	43%	

Note: All figures in 2004 Rupees unless noted otherwise

(Hibbs 2003). At the end of the economic lifetime of the breeder reactor, the value of the plutonium recovered is accounted for at the same cost at which it is assumed to be obtained from KARP.

The total construction cost of the PFBR is estimated at Rs. 34.92 billion (Rs. 28.08 billion in 2004 Rupees). The overnight unit cost is Rs 56,163/kW ($1,276.45 in 2004 Dollars) and is lower than the corresponding figure of Rs 60,335/kW for Kaiga III and IV. This is in contrast with experiences around the world that suggest that breeder reactors are much more expensive than water moderated reactors; for light water reactors (LWR), a typical estimate of the cost difference is $200/kW (Bunn *et al.* 2003).[10] The PFBR's estimated construction cost is also much lower than estimates of breeder reactor construction costs elsewhere; the Nuclear Energy Agency (NEA) gives a range of $1,850–2,600/kWe (in 2000 dollars) or $2,000–2,800 (in 2004 dollars) for MOX-fuelled fast reactors (NEA 2002). Both for these technical reasons and the DAE's history of cost overruns at *all* the reactors it has constructed, it is fairly likely that the PFBR capital cost will be higher than this projected value. Nevertheless, we will be using this figure in our calculations.

In economic terms, the primary material requirement for the PFBR is plutonium. The PFBR design requires an initial inventory of 1.9 tons of plutonium in its core (IGCAR 2003). Based on a detailed model of the reactor, it has been estimated that at 75 per cent capacity factor, the PFBR requires 1,012 kg of plutonium every year for refuelling during equilibrium conditions (Glaser and Ramana 2007). As mentioned earlier, the plutonium for the initial core and the first few reloads will have to come from KARP, whose production cost is Rs. 6,883/g. Because of the higher plutonium content of the PFBR spent fuel, the unit cost of subsequent plutonium requirements would be lower; we estimate it to be about Rs. 1,740/g (Suchitra and Ramana in preparation).

The plutonium is converted to MOX (mixed oxide) fuel by mixing with uranium. Fuel fabrication is to be done at the Advanced Fuel Fabrication Facility at Tarapur and the Fast Reactor Fuel Cycle Facility (FRFCF) at Kalpakkam (Balu, Purushotham, and Kakodkar 1998; AERB 2006). Unfortunately, no cost estimates are publicly available. Therefore, we follow the Nuclear Energy Agency and assume that the cost of fabricating driver fuel is $1,400/kg (in 2000 dollars) and the cost of fabricating (radial) blanket fuel is $500/kg (NEA 2002). In 2004 dollars, these are $1,512/kg and $540 respectively.

The corresponding cost figures for the PHWR have been discussed earlier. However, in this case, we use a uranium cost of $200/kg and a fabrication cost for uranium fuel of $200/kg.[11] Table 14.3 shows our preliminary results for the difference in the levelised cost, at a real discount rate of five per cent, of producing electricity at the PFBR and at Kaiga III and IV (Suchitra and Ramana in preparation).

The economics of the PFBR will be key to the future of breeder reactors in India. The DAE has argued that the 'primary objective of the PFBR is to demonstrate the techno-economic viability of fast breeder reactors on an industrial scale' (Chetal *et al.* 2006). Our results show that the PFBR will not be viable, even at the projected costs and for an optimistic capacity factor. If these assumptions do not hold, then its economic viability will be further reduced – for example, the time and cost overruns that are typical of the DAE might well occur with the PFBR and its associated facilities too.

As Table 14.4 shows, breeder reactors across the world have operated with relatively low lifetime capacity factors. If the PFBR experience were to be similar, a capacity factor of 50 per cent might be more plausible, and this would result in a levelised cost of Rs. 3.14 per unit, 89 per cent more expensive than PHWRs.

As mentioned earlier, the main rationale offered for the pursuit of expensive breeders is the shortage of uranium. We examine this by increasing the price of uranium from $200/kg to the 'crossover value' where breeders become competitive. For the optimistic base case, with a PFBR capacity factor of 80 per cent, the levelised costs of electricity from the PFBR and PHWR are equal at a uranium price of $860/kg. At a PFBR capacity factor of 50 per cent, the crossover price is $1,570/kg.

These prices are much higher than current values and significantly larger quantities of uranium will be available at these prices. The distribution of uranium among the major geological reservoirs in the earth's crust corresponds to a roughly

Table 14.4 Performance of breeder reactors

	PFR	*BN-600*	*Phenix*	*Superphenix*
Date of Construction Start	01-Jan-66	01-Jan-69	01-Nov-68	13-Dec-76
Date of First Criticality:	01-Mar-74	26-Feb-80	31-Aug-73	07-Sep-85
Date of Grid Connection:	10-Jan-75	08-Apr-80	13-Dec-73	14-Jan-86
Lifetime capacity factor	20.57%	71.51%	33.72%	6.6%

three-hundredfold increase in the estimated amount of recoverable uranium for every tenfold decrease in the ore grade (Deffeyes and MacGregor 1980). Based on this, and assuming that mining costs are inversely proportional to ore grade, one can surmise that the available uranium at costs less than \$860/kg and \$1,570/kg are 38 and 172 times current reserves respectively. This is an underestimate, because it ignores the general trends of reduced mining costs due to learning and improved technology (Schneider and Sailor 2005). In any case, India should have sufficient uranium to fuel PHWRs for decades, with no reprocessing and breeder reactors.

The preceding discussion is not meant to imply that instead of a breeder programme, a large-scale expansion of nuclear power based on heavy water reactors using high cost uranium is recommended. That would imply widespread uranium mining using ores of poor quality. As we describe in the next section, that has significant environmental and public health impacts. The point of our discussion of uranium availability is to demonstrate that the DAE has not undertaken the most elementary economic analysis necessary to justify the breeder programme.

Environmental impacts

Nuclear power brings other concerns. Different stages of the nuclear fuel chain release large quantities of radioactive and other toxic materials into the biosphere. Thus, a nuclear future merely trades radioactive externalities for carbon emissions at best.

Uranium mining and milling

Uranium mining and milling or refining has often severely impacted the health of local communities around the world (Makhijani *et al.* 1995; Eichstaedt 1994; Gilles 1996). When uranium ore is extracted from under the ground, it has to be chemically processed to separate out the uranium. The wastes created in this process, usually called mill tailings, are produced in large quantities because the typical amount of uranium in the ore is about 0.1 per cent or less.[12] They are contaminated with toxic heavy metals, such as molybdenum, arsenic and vanadium, and with radioactive materials, principally thorium-230 and radium-226. The radium-226 decays into radon gas, which can spread to considerable distances through the air (Eisenbud and Gesell 1997). Epidemiologic studies of underground miners from around the world have conclusively shown that inhalation of radon increases the risk of lung cancer; there is supporting evidence from experimental studies of animals and from molecular and cellular studies (BEIR 1988, 1999; UNSCEAR 2000).

Many of the other non-gaseous contaminants can affect the health of populations through ground water and locally grown food products. At virtually all US uranium milling sites, tailings have contaminated the ground water; once contaminated, 'the chance of returning an aquifer to pre-mining water quality is minimal' according to the US Nuclear Regulatory Commission (Makhijani *et al.* 1995).

Uranium mining and milling have extracted a toll on public health the world over. This is best illustrated by the experience at Jaduguda area, where increased incidences of births of babies with congenital deformities and of other undesirable outcomes have been widely reported (Rahman and Basu 1999; Tiwari 1999; Sonowal and Jojo 2003; Gadekar, Shreekumar, and Gadekar in preparation). Increases in numbers of birth defects, stillbirths, and other adverse outcomes of pregnancy have been reported in the case of the Shiprock uranium mining area in the United States as well (Shields *et al.* 1992). Similarly an excessive risk of leukaemia mortality was observed in the vicinity of uranium mills in Spain (López-Abente, Aragonés and Pollán 2001).

Nuclear reactors: routine releases

Nuclear reactors routinely release radioactive elements during their operations. Gaseous wastes produced during routine operations are usually released into the atmosphere. Low level liquid wastes are also released routinely into nearby water bodies, such as the sea in the case of coastal reactors.

Studies of populations near nuclear reactors have revealed several health impacts. For example, a meta-analysis of 17 research papers covering 136 nuclear sites in the UK, Canada, France, the United States, Germany, Japan, and Spain established that leukemia rates in children are elevated near nuclear facilities (Baker and Hoel 2007). In India, a comparative health survey of villages located near the Rajasthan Atomic Power Station (RAPS) at Rawatbhata and villages about fifty kilometres away from the site observed that the villages near RAPS had statistically significant increases in, *inter alia*, the rates of congenital deformities, spontaneous abortions, still births and one-day deaths of newborn babies, and of solid tumours (Gadekar and Gadekar 1996). The nearby villages also had fewer electricity connections.

Highly radioactive waste

The irradiated spent fuel arising out of the nuclear reactor contains the largest quantities of radioactivity produced in the fuel cycle. As mentioned earlier, there are two ways that countries have dealt with (or plan to deal with) such spent fuel: direct disposal and reprocessing.

Reprocessing, in many ways, is the dirtiest part of the nuclear fuel cycle, producing large amounts of solid, liquid and gaseous radioactive waste.[13] This is because reprocessing essentially separates out the large amount of radioactive substances contained in the spent fuel into three waste streams: low level, intermediate level, and high level. The low level waste is released into the biosphere and is therefore a conduit for various fission products to potentially reach human beings.

Whether they practice direct disposal or reprocessing, all countries envision disposing of the most radioactive parts, spent fuel in the former case and high level waste in the latter, in geological repositories. The idea of a geological repository dates back to the 1950s but turning the idea into reality has not taken place. Even if one were to be built, since there is no way of rendering the wastes benign, they will continue to be harmful to humans and other forms of life for hundreds of thousands of years. They have to be isolated and monitored if they are not to cause radiation doses. This need for stewardship is unprecedented in human history. This is also clearly iniquitous in respect of future generations, since they would bear the consequences while we use the electricity generated by the reactors. Ethical dilemmas aside, no technology that generates such long-lived radioactive wastes can be considered environmentally sustainable.

Safety

Nuclear power also poses a risk to energy security because it is susceptible to catastrophic accidents. Chernobyl is the best known instance of such a disaster. It resulted in several thousand deaths and the contamination of tens of thousands of square kilometres of land with radioactive elements like cesium-137 (NEA 2002a; Chernobyl Forum 2005; Greenpeace 2006). Agriculture across large parts of Ukraine and Belarus had to be suspended; over a hundred thousand people were relocated; and the economy of Belarus was devastated. Such accidents can happen in other (non-reactor) facilities too. In 1957, a tank containing radioactive wastes from the Mayak reprocessing plant in the erstwhile Soviet Union exploded and contaminated 20,000 square kilometres. India, still a largely agriculture-dependent economy, can simply not afford the risk of such disasters.

It is often stated that safety issues have been adequately addressed after the Chernobyl accident. However, the basic features of nuclear reactors remain the same. It is a complex technology involving large quantities of radioactive materials where events can spin out of control in a very short time. In studying the safety of nuclear reactors and other hazardous technologies, sociologists and organization theorists have come to the pessimistic conclusion that serious accidents are inevitable with such complex high-technology systems (Perrow 1984; Sagan 1993). The character of these systems makes accidents a 'normal' part of their operation, regardless of the intent of their operators and other authorities.

There is an experiential basis for concern about such accidents within India. Practically all the nuclear reactors and other facilities associated with the nuclear fuel cycle operated by the DAE have had accidents of varying severity (Chanda 1999; Rethinaraj 1999). Organizationally, DAE has not demonstrated a good safety culture (Kumar and Ramana forthcoming).

Breeder reactors are even more unsafe. For technical reasons having to do with their cores not being in the most reactive configuration, they are susceptible to catastrophic accidents involving large and explosive energy releases and dispersal of radioactivity. The PFBR is particularly at risk because of various design choices including high feedback effects and a weak containment building (Kumar and Ramana Submitted). The DAE's grandiose future plans are based on breeders using metallic fuels, which have an even higher feedback effect (Riyas and Mohanakrishnan 2008), with a correspondingly greater risk of accidents.

Conclusion

Nuclear power in India, as elsewhere in the world, is expensive when compared to the alternatives. Breeder reactors will be even more costly. Given the emerging nature of the electricity sector that emphasizes economic competition as the basis for deciding on generation choices, nuclear power is unlikely to be a significant source of power. Nuclear power also comes with the risk of catastrophic accidents, an important reason for adverse public opinion around the world. Finally, the claims of nuclear energy being environment friendly that one hears increasingly often, especially in the context of climate change, are baseless. The nuclear fuel cycle is polluting, albeit in a different way from fossil fuels.

The nuclear establishment in India has long promised much and, in exchange for unstinted government support, delivered little.[14] As per the DAE's predictions, by 2000 there should have been 43,500 MW of nuclear generation capacity in the country. What was achieved by 2000 was only 2,720 MW. Even now, nuclear capacity is only 4,120 MW, less than three per cent of the installed electricity generation capacity. Even if the capacity were to expand to 20,000 MW by 2020, the current goal, nuclear power would contribute only eight to 10 per cent of total electricity generation capacity. Thus, compared to the public pronouncements on the importance of nuclear power, we see that its actual contribution is a fairly small component of our electricity supply.

The failures of the DAE to meet its targets are not because of lack of resources. Practically all governments have favoured nuclear energy and the DAE budgets have always been high. This trend intensified after the 1998 nuclear weapons' tests and since then the DAE's budget has increased from Rs. 18.4 billion in 1997–8 to Rs. 50.3 billion in 2006–7, i.e., more than doubled even in real terms. In comparison, the 2006–7 budget of the Renewable Energy Ministry, responsible for 10,400 MW, contributing to 7.07 per cent of installed electrical capacity, was about Rs. 3.87 billion. Even with this limited government support, renewable technologies like wind power have grown impressively and become much cheaper in the last decade; India has the fourth largest installed wind energy capacity in the world. While their contribution to actual electricity generated would be smaller because these are intermittent sources of power, they have much lower maintenance costs. Further, exploitation of most of these sources started in earnest only relatively recently and there is ample scope for improvement.

Increased investment in renewable sources of energy is clearly desirable. Owing to increased R&D investments and cumulative operational capacity, the capital costs of several renewable energy technologies have been declining. This trend is likely to continue because, unlike mature technologies like coal and nuclear power, renewable energy technologies can improve considerably. These technologies are also amenable to decentralized, community-based production and cause much less environmental damage than fossil fuels and nuclear energy. Increased reliance on renewable energy technologies and improvements in energy efficiency offer a basis for a robust strategy that would go far in addressing both the goals of energy security and reduced environmental impacts. Nuclear power will do neither.

Notes

1 Though our paper will discuss the impacts of nuclear power production, this is true for other sources of power as well. Plans for hydropower, for example, have simply mentioned that there is a potential of 150,000 MW of hydropower, but there is no analysis whatsoever on how much land is required or how many people are likely to be displaced. As long as electricity is produced at competitive prices, it apparently does not matter if the inhabitants of the Narmada valley are evicted or if there is large scale contamination of land with fly ash.

2 This formulation simply mirrors the DAE's argument: 'The use of thorium in FBR in the third stage makes it a much larger resource (1,500 billion tonnes of coal equivalent) than the combined coal, oil and gas resources. Thus FBR provides long term energy security utilizing the indigenous uranium and thorium reserves' (IGCAR 2003).

3 As the psychologist Robert Jay Lifton put it, 'The central existential fact of the nuclear age is *vulnerability*' (Lifton and Falk 1982).

4 In this approach, all costs are discounted to some arbitrary but fixed reference date; the total cost reckoned at this reference point is the sum of the present values (PV) or future values (FV) of costs discounted to this date. For a description of this methodology see Brealey and Myers (2000).

5 Typical values chosen in costing electricity generation (or saving) technologies in India have ranged from 8 per cent to 10 per cent (real values) (for examples, see Shukla, Ghosh, and Garg 2003 and Nouni, Mullick, and Kandpal 2006).

6 The DAE also subsidizes the Nuclear Power Corporation by providing heavy water at a low lease rate, that too at a price that is much less than the cost of production (Muralidharan 1988; Ramana 2007).

7 It can also be used to make nuclear weapons.

8 The relatively scant amount of publicly-available data suggests that past performance of reprocessing plants in India has been mediocre. PREFRE, at Tarapur, operated at an average capacity factor of less than 25 per cent for over a decade (Hibbs 1995).

9 These uranium prices refer to long term contract prices, not spot prices that are typically reported and which tend to have high volatility due to short term demand-supply mismatches.

10 One reason for the increased cost is that breeder reactors use molten sodium as coolant. This brings with it several operational requirements, such as heating systems to keep the sodium molten at all times, and safety related requirements, such as extensive fire fighting equipment (Farmer 1984). Conversely, one reason for the relatively low cost of the PFBR is a design that compromises safety (Kumar and Ramana submitted).

11 This and a couple of other minor differences account for the slight difference in the calculated cost of electricity generation at Kaiga III and IV.

12 At 0.1 per cent ore grade, a thousand tons of uranium have to be extracted and processed to obtain one ton of uranium. At 0.01 per cent, it would be ten thousand tons.

13 For an estimate of the quantities of waste produced in India, see Ramana, Thomas and Varughese (2001).

14 This is arguably fortunate given the safety risks and environmental impacts associated with nuclear power.

References

AERB, 2006. Project Design Safety Committee for Fast Reactor Fuel Cycle Facility, IGCAR, Kalpakkam. (DSRC-FRFCF) Atomic Energy Regulatory Board 2006. Viewed 19 August, 2006 <http://www.aerb.gov.in/cgi-bin/committees/showcomm.asp?T=dsrcfrfcf.htm&V=DSRC-FRFCF>.

Baker, P. J. and Hoel, D., 2007. 'Meta-analysis of standardized incidence and mortality rates of childhood leukaemia in proximity to nuclear facilities.' *European Journal of Cancer Care*, 16, pp. 355–363.

Balu, K., Purushotham, D. S. C. and Kakodkar, A., 1998. 'Closing the fuel cycle – a superior option for India.' Paper read at Fuel Cycle Options for Light Water Reactors and Heavy Water Reactors, 28 April – 1 May 1998, Victoria, Canada.

Balu, K, and Ramanujam, A., 1999. 'Reprocessing and recycle of U/Pu – a safer option for optimum utilization of resources in the nuclear fuel cycle.' In Pushparaja, R. K. K., Sangurdekar, P. R. and Kurien T. (eds), *Radiation protection in nuclear fuel cycle: control of occupational and public exposures*. Kakrapar: Indian Association for Radiation Protection, pp. 57–62.

BEIR, 1998. *Health risks of radon and other internally deposited alpha-emitters (BEIR IV), 1988*. Washington: Committee on Biological Effects of Ionizing Radiation, National Research Council, National Academy Press.

BEIR, 1999. *Health effects of exposure to radon (BEIR VI)*. Washington: Committee on Biological Effects of Ionizing Radiation, National Research Council, National Academy Press.

Bhabha, H. J. and Prasad, N. B., 1958. 'A study of the contribution of atomic energy to a power programme in India.' Paper read at Second United Nations International Conference on the Peaceful Uses of Atomic Energy, 1958, Geneva.

Brealey, R. A. and Myers, S. C., 2000. *Principles of corporate finance*, 6th ed. Boston: Irwin McGraw-Hill.

Bunn, M., Fetter, S., Holdren, J. P. and van der Zwaan B., 2003. *The economics of reprocessing vs. direct disposal of spent nuclear fuel*. Cambridge: Harvard University.

Bunn, M., Holdren, J. P., Fetter, S. and van der Zwaan, B., 2005., 'The economics of reprocessing versus direct disposal of spent nuclear fuel.' *Nuclear Technology*, 150, pp. 209–230.

CERC, 2006. Annual Escalation Rates for Bid Evaluation: Central Electricity Regulatory Commission Notification Eco 2/2006-CERC, October 26, 2006. Viewed 9 November, 2006.

Chanda, N., 1999. 'The perils of power.' *Far Eastern Economic Review*, 4 February 1999.

Chernobyl Forum, 2005. Chernobyl's legacy: health, environmental and socio-economic impacts and recommendations to the Governments of Belarus, the Russian Federation and Ukraine. Vienna: International Atomic Energy Agency.

Chetal, S. C., Balasubramaniyan, V., Chellapandi, P., Mohanakrishnan, P., Puthiyavinayagam, P., Pillai, C. P., Raghupathy, S., Shanmugham, T. K. and Pillai, C. S., 2006. 'The design of the prototype fast breeder reactor.' *Nuclear Engineering and Design*, 236, pp. 852–860.

Chowdhary, S. K., 1998. 'Coal in India.' Presentation at the 17th World Energy Congress, September 13–18, 1998, Houston.

Deffeyes, K. S. and MacGregor, I. D., 1980. 'World uranium resources.' *Scientific American*, January 1980, pp. 66–76.

Deutch, J., Moniz, E. J., Ansolabehere, S., Driscoll, M., Gray, P. E., Holdren, J. P., Joskow, P. L., Lester, R. K. and Todreas, N. E., 2003. *The future of nuclear power: an interdisciplinary MIT study*. Cambridge, MA: Massachusetts Institute of Technology.

Dey, P. K., 2003. 'Spent fuel reprocessing: an overview.' Paper read at Nuclear Fuel Cycle Technologies: Closing the Fuel Cycle, December 17–19, 2003 at Kalpakkam.

Eichstaedt, P. H., 1994. *If you poison us: uranium and Native Americans*. Santa Fe, N M: Red Crane Books.

Eisenbud, M., and Gesell, T. F., 1997. *Environmental radioactivity: from natural, industrial, and military sources*. 4th ed. San Diego: Academic Press.

Farmer, A., 1984. 'Assessing the economics of the liquid metal fast breeder reactor.' In Brookes, L. G. and Motamen, H. (eds.) *The economics of nuclear energy*. London: Chapman and Hall.

Gadekar, S. and Gadekar, S. N., 1996. 'Rawatbhata.' In Gaur, V. (ed), *Nuclear energy and public safety*. New Delhi: Indian National Trust for Art and Cultural Heritage.

Gadekar, Surendra, Shreekumar and Gadekar, Sanghamitra. (In preparation). *Health impacts of uranium mining in Jaduguda, India.*

Gilles, C., 1996. 'No one ever told us: Native Americans and the great uranium experiment.' In Bryne, J. and Hoffman, S. M. (eds.) *Governing the atom: the politics of risk.* New Brunswick: Transaction Publishers.

Glaser, A., and Ramana, M. V., 2007. 'Weapon-grade plutonium production potential in the Indian prototype fast breeder reactor.' *Science and Global Security*, 15, pp. 85–105.

Greenpeace, 2006. The Chernobyl catastrophe: consequences on human health. Amsterdam: Greenpeace.

Hibbs, M., 1995. 'Tarapur-2 to join twin BWR in burning PHWR plutonium.' *Nuclear Fuel*, 20(20), p.18.

Hibbs, M., 2003. 'DAE reprocessing program remains modest in scope.' *Nuclear Fuel* 28(8), p. 18.

IGCAR, 2003. 'Design of prototype fast breeder reactor.' Indira Gandhi Centre for Atomic Research 2003., Viewed 10 March, 2006 <www.igcar.ernet.in/broucher/design.pdf>.

Kakodkar, A., 2006. 'Securing our emerging energy needs: what nuclear energy can do.' *Energy Security Insights*, 1(1), pp. 4–7.

Kumar, A. and Ramana, M. V. (Forthcoming). 'Nuclear safety in India: theoretical perspectives and empirical evidence.'

——Submitted. 'Compromising safety: design choices and severe accident possibilities in India's prototype fast breeder reactor.'

Lifton, R. J. and Falk, R., 1982. Indefensible weapons: the political and psychological case against nuclearism. New York: Basic Books.

López-Abente, G., Aragonés, N. and Pollán, M., 2001. 'Solid-tumor mortality in the vicinity of uranium cycle facilities and nuclear power plants in Spain.' *Environmental Health Perspectives*, 109(7), pp. 721–729.

Makhijani, A., Hu, H. and Yih, K., 1995. *Nuclear wastelands: a global guide to nuclear weapons production and its health and environmental effects.* Cambridge, Mass: MIT Press.

Muralidharan, S., 1988. 'Birth of nuclear power corporation.' *Economic and Political Weekly*, 23(5), pp. 190–192.

National Research Council, 1996. *Nuclear wastes: technologies for separations and transmutation.* Washington, DC: National Academy Press.

NEA, 2002. Accelerator-driven systems (ADS) and fast reactors (FR) in advanced nuclear fuel cycles. Paris: Nuclear Energy Agency, OECD.

NEA, 2002a. Chernobyl: assessment of radiological and health impacts (2002 update of Chernobyl: ten years on.) Paris: OECD Nuclear Energy Agency.

Nema, A. K., 1999. 'Nuclear generation cost in India.' *Nu-Power*, 13(1).

Nouni, M. R., Mullick, S. C. and Kandpal, T. C., 2006. 'Photovoltaic projects for decentralized power supply in India: a financial evaluation.' *Energy Policy*, 34(18), pp. 3727–38.

Perrow, C., 1984. Normal accidents: living with high risk technologies. New York: Basic Books.

Planning Commission, 2006. *Integrated energy policy: Report of the Expert Committee.* New Delhi: Planning Commission, Government of India.

Prasad, A. N., 1996. 'Spent fuel reprocessing: a technology at cross roads.' In *Seminar on Indian Reprocessing, Kalpakkam,* pp 13–18, Project KARP, BARC Facilities.

Rahman Aziz ur and Basu.,1999. 'Living in death shadow', *Sunday*, 4 April.

Ramana, M. V., 2007., 'Economics of nuclear power: subsidies and competitiveness.' *Economic and Political Weekly*, 42(2), pp. 169–171.

Ramana, M. V., 2007. 'Heavy subsidies in heavy water: economics of nuclear power in India.' *Economic and Political Weekly*, 42(34), pp. 3483–90.

Ramana, M. V., D'Sa, A. and Reddy, A. K. N., 2005. 'Economics of nuclear power from heavy water reactors.' *Economic and Political Weekly*, 40(17), pp. 1763–73.

Ramana, M. V. and Suchitra, J. Y. 2007. 'Costing plutonium: economics of reprocessing in India.' *International Journal of Global Energy Issues* 27(4), pp. 454–471.

Ramana, M. V., Thomas, D. G. and Varughese, S., 2001. 'Estimating nuclear waste production in India.' *Current Science* 81(11), pp. 1458–62.

Rethinaraj, T. S. G., 1999. 'In the comfort of secrecy.' *Bulletin of the Atomic Scientists*, 55(6), pp. 52–57.

Riyas, A. and Mohanakrishnan, P., 2008. 'Studies on physics parameters of metal (U-Pu-Zr) fuelled FBR cores.' *Annals of Nuclear Energy*, 35(1), pp. 87–92.

Sagan, S., 1993. *The limits of safety: organizations, accidents and nuclear weapons*. Princeton: Princeton University Press.

Schneider, E. and Sailor, W. C., 2005. *Long term uranium supply issues and estimates*. Washington, DC: Los Alamos National Laboratory, US Department of Energy.

Sethna, H., 1979, 'India's atomic energy programme – past and future.' *IAEA Bulletin*, 21(5), pp. 2–11.

Shields, L. M., Wiese, W. H., Skipper, B. J., Charley, B. and Benally, L., 1992. 'Navajo birth outcomes in the Shiprock uranium mining area.' *Health Physics*, 63(5), pp. 542–51.

Shukla, P. R., Ghosh, D. and Garg, A., 2003. 'Future energy trends and GHG emissions for India.' In Toman, M. A., Chakravaraty, U. and Gupta, S. (eds.) *Global climate change: perspectives on economics and policy from a developing country*. Washington DC: Resources for the Future.

Sonowal, C. J. and Jojo, S. K., 2003. 'Radiation and tribal health in Jadugoda: the contention between science and sufferings.' *Studies of Tribes and Tribals*, 1(2), pp. 111–26.

Srinivasan, M. R., 1985. 'The Indian nuclear power programme.' Paper presented at Indo-French Seminar on Nuclear Energy, Bombay, 1–4 April 1985.

Suchitra, J. Y. and Ramana, M. V. (In preparation.) 'The costs of power: plutonium and the economics of India's prototype fast breeder reactor.'

Thakur, S. and Chaurasia, B. P., 2005. *Cost effectiveness of electricity generating technologies*. Mumbai: Nuclear Power Corporation.

Tiwari, M., 1999. 'A deformed existence.' *Down to Earth*, 15 June.

UNSCEAR, 2000. *Sources and effects of ionizing radiation: UNSCEAR 2000 report to the General Assembly, with scientific annexes*. New York: United Nations Scientific Committee on the Effects of Atomic Radiation, United Nations.

Von Hippel, F. and Jones, S., 1997. 'The slow death of the fast breeder.' *Bulletin of the Atomic Scientists*, 53(5), pp. 46–51.

15 India's energy security landscape

Joining the dots and looking ahead

Ligia Noronha

The current debate on India's energy security is framed by three key issues: energy needs related to growth, the stark energy poverty in the country, and the implications of its energy paths to future contribution to climate change. They all involve some form of risk: energy poverty has environmental health implications,[1] involves special burdens on women and children, and results in a lack of educational facilities and jobs and reduced opportunities. Energy and growth linkages can lead to possible competition over securing resources and even traditional security conflicts. This is especially pertinent for large growing economies like India and China, and large developed economies such as the United States that depend on external energy sources. India's energy mix, as that of most other countries, is dominated by fossil fuels, and global climate risks are linked with how a country chooses to fuel its economy in the future. It is within this linked energy-environment-development context that the understanding of, and ways to address, India's energy security concerns need to be located.

India's development objectives require the Indian economy to grow at eight per cent per annum over the next two and a half decades; the estimates of the Indian Planning Commission suggest that this requires the commercial energy sector to grow at –five to six per cent p.a. to 2031 over the 2003–04 levels, and in a business as usual situation this will mean that energy supplies will have to be 7.5 times current amounts. The fuel mix is very much based on fossil fuels and over this period India is expected to become 59 per cent import-dependent for its commercial energy requirements. With rising oil and coal prices, a rising resource nationalism, increased risks and threats to energy infrastructure, and the implications of the effect of a large increase in fossil fuel use on carbon emissions, India's energy concerns have been growing acutely, and with it the 'securitization' of the debate around it. From 'availability' and 'affordability', the need for 'secure' energy resources has become central to maintaining and achieving economic dynamism; the objective of energy independence is becoming very important, but strategies pursued to secure this 'independence' – equity investments, bilateral deals and new energy ties, new thinking on and alignments for nuclear energy – have tended to create regional and global concerns about their implications. This volume has sought to highlight some of the key concerns that relate to India's energy security, concerns relating to access, fuel choices, fuel sourcing, consumption, and environmental sustainability and how these are being addressed or need to be addressed. This concluding chapter briefly brings together some of the key messages that have emerged from the essays in this volume, and reflects on the way forward.

Key messages

Section I of this volume concentrated on mapping India's energy security concerns and energy policy making. It is evident that given India's energy requirements, its energy policy must focus on exploiting a variety of energy sources, while at the same time dealing with social and environmental inequities and damages. Within this already complicated framework, it is by no means easy to ensure that the national energy policy pays heed not just to ensuring the availability of sustainable energy, but also assuages concerns about security and stability. Thus there are a number of dimensions to the problem. They include internal factors, such as consumption patterns, a flawed domestic policy and infrastructural choices, and energy poverty; and an external dimension – a rising import bill, the geopolitics of global energy trade and military vulnerabilities. Of these two, it is the external dimension of energy policy that is most closely identified with energy security, although it is by no means a complete description of the issue, which is why this volume sought to focus on understanding the context of India's energy security concerns in Section I.

Sudarshan and Noronha, in the opening chapter, focussed on India's fossil fuel path dependence, an energy path which in general is not dissimilar to that followed by the developed world. This chapter sought to ask why India has struggled to implement change, even where attempts have been made, and to explain the struggle through the use of new institutional economics, and with a focus on four constraints – technology, relative prices, beliefs and perceptions, and institutions and organizations.

Sethi discusses the key energy challenge India faces, which is to decouple economic growth (which is key to address India's poverty) from growth in energy consumption while ensuring universal access to lifeline levels of energy consumption. He highlights the three most important challenges for India: The need to access more energy at 'affordable' prices; the growing pressure on its incremental contribution to greenhouse gases from the use of coal resources; and, finally, the challenges posed by an inadequate domestic energy infrastructure. He provides arguments for several of the choices India is making in its energy strategies, and explains why some are preferred over others. India's energy strategies involve developing domestic energy sources, finding new sources, and energy efficiency measures. The balance of India's requirements is imported or sourced overseas in terms of equity investments. Sethi argues strongly for choices that have inclusive growth results and that are also environmentally sustainable. As one of the key architects of India's Integrated Energy Policy of 2006, Sethi's comments provide pointers to understanding energy choices and the limits of India's room to manoeuvre on this front.

India's energy choices and strategies have become the centre of regional and global attention for a number of reasons. One is that, along with a somewhat unfair hyphenation with China, India's growing energy demand is said to be exerting a new pressure on world supplies. On the other hand, its low per capita energy consumption, a fifth of the world's average, implies that a large part of its population is still starved of energy, which influences its development options and affects its human condition. A second reason is that a large part of India's fuel mix is carbon based. India's development needs and current energy choices require the global community to provide it with a greater share of the shrinking ecological space, as defined by the limits of carbon absorptive capacity. Alternatively, it needs assistance, financial and technical, to move to low carbon paths. But given India's emerging global presence, there is a concern that such assistance, while it may address a global public 'bad' such as climate change, may work to the competitive disadvantage of the developed world. Third, there is a concern among OECD countries that India's overseas energy initiatives reduce the ability of OECD countries to further own foreign policy agenda on democracy and governance, and hurt other international energy companies as they perceive Indian (and Chinese) companies competing unfairly. I return to this point later in the chapter. Fourth, an issue of domestic rather than international concern; is why clean energy sources such as hydropower, for example, have not taken off in India, despite their potential. I would argue that this is primarily because inadequate attention has been paid to equitable benefit-sharing mechanisms, land-use compensation, and the rehabilitation aspects of such developments. Compensation for land that is used up for energy-resource development must go beyond one-time payments to project-affected people and should include employment opportunities, share in revenue streams, and lease/rental value for the land, with an escalation factor.[2]

Gupta and Sudarshan have focused on a detailed review of the literature on energy and poverty linkages and provide an analysis of the energy use patterns and transitions for cooking and lighting in India. They argue that while availability alone is not enough to motivate shifts in energy use, enabling widespread access to modern alternatives is crucial if transitions are to work. They point out that when fuel supply and distribution systems are controlled by the state, there tend to be inequities in distribution, black markets, and corruption. Enabling easy access, they argue, is making it possible for poor households to purchase small amounts of energy at a time to make it affordable. As far as electricity is concerned, low-ampere connections need to be made available. Thus ensuring the availability of electrical connections that take into account the needs of the consumer is important, and it is not necessarily the case that one size should fit all.

The chapter also highlights a long-standing debate in India that while, in principle, subsidies can increase the ability of the poor to access modern energy sources, broad based subsidies perversely tend to benefit the better off much more than the truly poor, because of the greater consumption and greater ease of access of those who are better off. The use of smart cards to target subsidies has also been much talked about but is yet to be implemented in India.

Section II puts Indian energy needs in a global context and examines India's challenges and strategies in the light of rising energy prices, market power and geopolitical developments. This discussion helps place in context the vulnerabilities that India faces, purely as a consequence of being a heavily import dependent nation (especially with regards to oil). It also makes the point that the oil market in itself, even aside from geopolitical concerns, has players with significant market power and consequently is unlikely to reflect the characteristics of a competitive system. Nanda argues that optimising the energy-use mix in the country on the basis of future scenarios in different energy products in the global market should be, and is, a policy imperative. He points out that India currently has a high dependence on oil and coal but very little use of gas, but given that oil has always been a problem for India, and that higher imports of coal in the future will also have their own problems, India must explore the option of increasing the use of gas and renewable forms of energy. A greater engagement with natural gas will undoubtedly require India to deal with the geopolitical challenges surrounding natural gas development, production, and distribution at the international level, with a need for attention to the development of markets, infrastructure, and regulation at the domestic level.

Ahmad, through a broad discussion of issues relating to the location of key oil and gas resources in Western and Central Asia, and through an argument of why these regions are key to India's oil and gas security, highlights a number of different interests and alliances that are being pursued in Asia. In his chapter, he puts forward the view that India perceives the quest for energy security as requiring a collaborative approach rather than one defined in terms of competition and conflict. He points out that this quest needs to engage and build relations with other consumers and producers. His chapter provides various examples of India's energy, especially oil, diplomacy. Some of these initiatives and ties are, as mentioned before, raising a concern among Western nations that they undermine their own efforts at improving governance and human rights in those countries.The counter to this view is, of course, that a large number of international companies have been doing business with these countries for a long time, and this new refrain is really just a concern that they may be losing ground to the new competitors. I have also argued in several fora on previous occasions that these ties, while seeming expedient in the short run, can lead to increased stakes in greater stability and peace in regions currently troubled.

The transnational natural gas pipeline story in India is one of failures rather than successes. Batra focused on a discussion of the four pipelines that have been continually in and out of the public and political gaze. His key observation is that India has failed to tie up a single pipeline project in its neighbourhood, despite being surrounded by countries with natural gas. He argues that geopolitics, security concerns about pipelines and assurance and pricing of gas supply have been key factors in this failure. It is abundantly clear that if these or any other transnational pipeline projects are to take off, there is need for greater investor confidence, trust and cooperation. Absence of trust, and the existence of sanctions, embargoes, and threats, tends to result in under-investment in energy-resource development through reduced capital flows. The economic impacts of such political choices are then felt keenly through the reduced availability of energy supplies.

The issue of cooperation or conflict with China on energy-securing issues has been the subject of considerable conjecture and debate over the last three years. In many ways, both countries have similar energy profiles and concerns and similar internal and external energy-securing strategies. The key difference seems to be in the intensity with which these are pursued, China being far more intense than India in its energy pursuits. The international debate has focussed considerably on the potential for competition and conflict between these two emerging economies as they search for resources. There have indeed been some instances of competition, wherein China has taken the prize. In her chapter, Mahalingam, however, focuses on what seems to make more long-term sense for India and China – energy collaborations in areas where this makes sense. She argues for planned initiatives and a movement from the general to the specific in order to get them off the ground.

Khurana highlights India's very high dependence on maritime routes for the transport of imported energy and analyses the geopolitical reasons why they, rather than the overland route, will continue to remain important in future. Given this, he discusses the various kinds of potential threats – state and non state – to these energy 'lifelines', and emerging security issues, which can result in an increased vulnerability for India, in the context of geostrategic imperatives. He argues for appropriate policy responses towards secure 'maritime logistics' of energy resources.

Raja Mohan explores the larger conceptual challenges that Indian foreign policy must confront in ensuring energy security for the nation. He does this by examining some key elements of India's energy strategy to assess the implications these could have for India's traditional foreign policy, and what might need to change if they were to be successful. He argues that energy security is not just an additional demand on India's foreign policy, but that, given its scale and scope it requires a fundamental change in the principles that have guided this policy. The transition towards a more responsible foreign policy, he suggests is needed, at once committed to securing India's national interests as well as contributing to collective goods.

Section III has deliberately focussed on issues of energy consumption with a view to assessing what can be done to address consumption proactively. While India has an acute energy poverty problem, there is an emerging class that is beginning to adopt very lavish, energy consuming lifestyles. It is to this emerging concern that Das Gupta turns in her chapter, and provides an early warning of what such lifestyles could mean for both energy security and climate change. While her chapter focuses on the need for individual and household behaviour to avoid the trodden path of overconsumption, the chapter by Dadhich reports on an earlier TERI study which, through modelling scenarios, highlights the implications of technology choices for increased energy efficiency and cleaner energy use more widely across sectors. Mathur *et al.* take this a step forward and examine the policies and government initiatives that have been put in place in India to influence and incentivize more energy-efficient behaviour and consumption choices. Their paper also discusses the observed constraints to improved behaviour and to the uptake of cleaner technologies.

An important aspect of energy security that has not been covered in much detail, almost not at all in this volume, is energy pricing. All four energy policy concerns – access, security, efficiency, and environmental sustainability – demand a rational approach to energy pricing in order to provide right signals to producers, consumers and investors. As long as India was insulated from the global market, this could to some extent be tolerable, although the price rise of the 1970s pressured India enormously given the burden on balance of payments from oil imports. But the projected increase in dependence on imports of oil, gas and coal, as well as the need for creative incentives to move to a low carbon economy, makes energy pricing a key aspect of energy decision-making. Whether it is to incentivise change, or to create more oil-independent lifestyles, pricing is a key instrument to leverage change in consumption patterns and technology choices. Energy pricing in India, however, has been particularly susceptible to administrative fiat and the calculus of politics. And oil prices have always been a particular target of policy attention to insulate the economy from external forces, subsidize its use by some groups, as well as to provide revenues for the government. Take the recent handling of rising international oil prices by the Indian government: with international oil prices at USD135/bbl and having risen sharply over the last two years, product prices at the pump had gone up just twice. In part this was because the Indian consumer had already for a long time been paying a high price for petrol and diesel because of the high share of taxes (excise duties, education cesses and State sales tax) in the price build up, and so in many ways it was unfair to pass on the higher international oil prices. But in 2008, because of this control on prices, petroleum product prices at the pump, despite the taxes, have been lower than they should be if the right signals to conserve and reduce the use of oil were to be given. The burden of the subsidy was borne by oil marketing companies and the exchequer and future taxpayers, as it was financed through under-recoveries by oil companies and the issue of oil bonds. In very early June 2008, before the latest revision in prices and the reduction in taxes, oil marketing companies were losing Rs.13.8/litre of petrol and Rs.25/litre of diesel due to domestic pricing polices.[3] In the case of LPG and kerosene, used for cooking and lighting, the price differential between the price paid for the Indian basket of crude and the subsidized price has also been borne by oil marketing companies, and with rising crude oil prices the under-recovery on these fuels has been increasing (Misra *et al.* 2008). The stark truth is that these subsidies, ostensibly introduced to help the poor, are not reaching the poor, and are in fact being diverted to service the better-off classes. TERI has been mooting the use of smart cards to target such subsidies to the intended groups (Misra *et al.* 2005), and the Integrated Energy Policy, released in August 2006, mentions smart cards as an important option for the provision of targeted subsidy delivery to the needy. But whether it is fuel for transport or for cooking and lighting, the issue of pricing and targeted subsidies are caught in non decisions and politics. Few attempts exist to rationalize the system or to address if we should in fact be subsidizing the use of oil or diverting these funds into investments of renewable energy. Meanwhile the concern is that wrong signals are being sent out and wrong incentives created in terms of influencing choices about energy consuming and producing. The high subsidy that is given to diesel is in fact leading to what has been termed the 'dieselization of the Indian economy'[4].

In Section IV of the volume we focus on the nuclear debate in India. We deliberately chose two polar papers in order to highlight that, in the Indian case, the 'jury is still out' on the role that nuclear energy should play in India's energy future.[5] It is evident that India does not have many choices as far as fuels go; its richest energy source may well be solar power. But until this is well and truly established, the country needs to use every energy source that it can work with. The power projections over the next 30 years make this an important option. Grover's chapter makes a strong case for the nuclear energy source and clearly discusses its development in the Indian context. He also makes the point that the economics of nuclear power generation are complicated to determine, and that there are grounds to believe that it makes sense in an Indian context. The argument that nuclear energy is a cleaner fuel than fossil fuels in terms of carbon emissions has contributed to its increased acceptability, but there are still many grave unknowns and uncertainties – economics, the environmental and human impact of possible disasters, discussed in the chapter by Ramana and Suchitra – which make this energy source difficult to bat for unreservedly. The authors argue that the environmental costs of nuclear power are high, and the economics uncertain. They point out that these concerns should not be forgotten simply because nuclear energy happens to be freer of carbon emissions.

Growth of nuclear energy in India is constrained by India's uranium reserves. This was the key reason the Indian nuclear establishment opted for a three-stage nuclear programme very early on, as discussed in Grover's chapter. But thorium technology is 30 to 40 years away and to run it there is a need to access the global market for uranium as the cycles are interlinked, and the third phase is possible only after 50GW of the first and second phase of the programme are established. But since India has not signed the Non-Proliferation Treaty, it has been unable to import uranium. The recent waiver by the Nuclear Suppliers Group will open international opportunities for civil nuclear trade. Without such a waiver, India's nuclear energy, based on just indigenous resources, could increase from 3.31 GW in 2006, to 6.78 GW by 2010, and to 21.18 GW by 2020. It would stay at this beyond 2021, given limited domestic uranium reserves and no imports. However, with the waiver and a successful implementation of the Indo–US nuclear collaboration, DAE expectations are that nuclear energy generation capacity can go up to 70 GW by 2031–32 and to 275 GW by 2052. If the third stage becomes operational, using FBR technology and thorium as fuel the estimated potential from nuclear is 530 GW. Much, however, depends on the fine print of the nuclear energy deal and its interpretation and acceptance by the concerned parties, as well as internationally.

The Indo–US nuclear energy collaboration has served to highlight a basic divergence of ideologies in the country, particularly with regard to India's place in the world and its relationship with the United States. However, the specific political debate on the nuclear agreement in India and the United States is an ongoing dynamic process, one which has seen, and is continuing to see, rapid developments. For that reason it is hard to discuss the specifics of that agreement in a volume such as this one, though the ramifications of this experience (whatever the eventual outcome of the deal) will certainly have important implications for India's future energy policy. Raja Mohan in Section II does provide a brief top-level discussion of some of the lessons that can be drawn from the controversy over the nuclear agreement, as well as some discussion on what it means in the context of India's foreign policy and international energy policy. But it is clear that it is politics, not economics, which will drive decisions on the nuclear option, as is the case in other countries.[6]

Looking ahead

As the chapters in this volume highlight, addressing energy security is complex, enmeshed as it is in domestic and global politics. The key challenges that India faces, and will face include the following.

- Increasing economic growth while trying to keep growth in energy demand down. This is especially challenging due to the transition from biomass to fossil fuels, the growth of motorized private transport, and rising incomes, aspirations and changing lifestyles.
- Severe energy inequities and the likely persistence of a significant share of traditional fuels in the primary energy mix for at least another decade.
- The 'lock-in' characteristics of India's fuel mix, where coal and oil remain the dominant energy sources across a variety of scenarios; and supply, technology and cost constraints restrict the adoption of alternatives such as natural gas, renewables and nuclear energy.
- High current and projected import dependencies, particularly on coal and oil, and consequent geopolitical and trade implications and increased import bills, given rising fuel prices.
- Global and local environmental concerns and the possibility of constraints placed on energy choices as part of the international climate change negotiations process.

Addressing such challenges needs a global-national-local perspective. India's energy security will increasingly depend on action along three fronts: the diversification of energy supplies, the 'securing' of energy supplies, and managing energy consumption. The *diversification of energy supplies* in turn implies the following: (1) a search for new markets, (2) search for secure, stable energy partners and routes of transportation and (3) search for affordable energy supplies, through oil equity, joint ventures, production-sharing arrangements and so on. The *'securing' of energy supplies* will need to involve: (1) a pragmatic and skilful foreign policy that is capable of leveraging the national interest within a complex regional and international geopolitical environment, (2) a long-term view of the security of energy supplies in terms of the impact of resource-based trade and investment on the country of origin, and (3) a long-term view of energy demand and production in light of environmental concerns such as climate change. *Managing energy consumption* will involve: (1) enhanced efficiency in energy consumption so that a certain level of energy services is met by consuming less energy. This is applicable to 'islands' of high energy intensity such as industry, transport and high-income households; (2) addressing social inequities in energy access resulting from differences in physical access as well as affordability; (3) reduction in the environmental and carbon footprints of energy consumption through a judicious choice of clean fuels and improved energy use technologies.

Given the transnational nature of the challenges, multilateral efforts are also needed on the issue of high and rising oil prices, nuclear-energy trade, increased cooperation on renewable energy sources; and towards more equitable global energy consumption that increases ecological space for less developed economies to provide energy to their populations. It is immediately obvious that addressing such issues globally also requires a competitive and healthy domestic energy sector that avoids price distortions and wrong signals to producers and consumers; flexible institutions that can respond to shocks and changes to the system; and an increasingly aware domestic consuming class.

Nations cannot hope to achieve energy security without entering into mutually beneficial and interdependent relations with other countries. As Yergin put it: '… it must be recognized that energy security does not stand by itself but is lodged in the larger relations among nations and how they interact with one another' (Yergin 2006). In our view, these relations have to be collaborative and without conflict. Securing energy, although seemingly an objective in the national interest, needs, in its achievement, the quest for improved relations with both consuming and producing nations. A quest for energy security that results in nations being pitted against each other and in competition, with elements where there is the potential for conflict, can only be a short-term vision that will result in zero-sum solutions. What are needed are collaboration and partnership, and the building of trust. There is an urgent need to understand the interdependence of energy systems and the complementary interests of energy producers and consumers, and the need for stability in markets and supplies. While energy security has hitherto been discussed from the perspective of importers and consumers, increasingly the debate is getting enlarged. It needs to get enlarged to include the exporters, as it is becoming evident that long-term security lies in recognizing and deepening the interdependence.

India needs to be part of, and central to, a new architecture for energy cooperation. This can be at various levels: bilateral, regional, and global; and can take many forms – the development of energy hubs, energy dialogues, joint stockpiling, joint research and development into new technologies, and into clean coal technologies. It also needs to be part of global initiatives to stabilize prices. Identification of energy hubs within the South Asian region can serve to not only integrate the South Asian region within, but also connect a principal member of the South Asian region to an external energy supplying country or region.[7]

The recent announcement of the setting up of a Solar Energy Commission along the lines of the Atomic Energy Commission as part of India's Climate Change Action Plan is very welcome, as it was never clear why India, with abundant solar resources, never gave the development of this technology the rightful place that it deserved. From the point of view of resource availability, access, avoidance of geopolitics, the ability to have distributed decentralized generation, and environmental sustainability, solar power is a natural solution to a large part of India's energy insecurity. There exist possibilities for countries to take up joint ventures wherein India could provide technology and expertise with regard to wind power generators. India may also provide technology and/or finance for wind installation in another country's territory, which could then feed in a proportion of the generated power to the Indian grid. India's lead in the manufacture of solar panels and photovoltaic cells too can provide the basis for collaborative ventures and trade. Surplus hydropower availability in the neighbourhood of Bhutan and Nepal makes them important contributors of hydro-electricity to India. These initiatives have an added value not only in that they help build neighbourhood relations and increase energy availability, but they also help address climate change concerns. This seems an opportune time to change the way we do energy business. This is a time when we need to go beyond state-centred views of energy security and find ways in which we can work with a more global perspective of energy security that also addresses climate concerns. A deep commitment is needed to more financing, access to clean technologies, greater international and national collective action, and changes in the way we live our lives.

Technology for cleaner development offers some room to manoeuvre, as we have seen, but the scale of the energy and climate problem is huge and the full extent of the future implications of climate change is still so uncertain that unless we change our consumption patterns, the future is at great risk for all of us. There is no better time or stronger arguments to get off the path of fossil fuel path than now. It is urgent therefore, that we engage with all the key sources of this dependence to effect a process of energy change.

Notes

1 For example, the WHO estimates that indoor smoke from solid fuels kills 1.6 million people a year. See http://www.who.int/indoorair/health_impacts/burden_global/en/index.html, viewed 25 September, 2007.
2 See 'Study on compensation to resource bearing States,' prepared by The Energy and Resources Institute for the Inter-State Council Secretariat, Ministry of Home Affairs, India, 2007.
3 'The dilemma called Oil' V Mehta, Indian Express, 3 June, 2008.
4 'The dilemma called Oil' V Mehta, Indian Express, 3 June, 2008.
5 See also Section III in Sharma and Noronha, 2008, *Energy, climate and security: the inter-linkages* Proceedings of the Second TERI-KAS Conference, KAS publication Series No. 19, New Delhi.
6 See the comments by Ashley Tellis in 'The Nuclear Comeback: An Option for Sustaining Global energy and Climate Security?' in Sharma and Noronha, 2008, op cit.
7 Sharma and Mahajan, inputs from research done under the Khemka project.

References

Misra, N., Chawla, R., Srivastava, L. and Pachauri, R. K., 2005. *Petroleum prices in India: balancing efficiency and equity.* New Delhi: TERI.

Misra, N., Chawla, R. and Srivastava, L., 2008. 'Towards effective petroleum subsidies.' *Energy Security Insights*, April.

Sharma, D. and Noronha, L., 2008. *Energy, climate and security: the inter-linkages.* Proceedings of the Second TERI-KAS Conference, KAS publication Series No. 19, New Delhi.

TERI, 2007. 'Compensation to resource bearing States: minerals, coal and hydropower.' [Report for the Inter State Council], New Delhi.

Yergin, D., 2006. 'Foreign Affairs', March/April.

Conversion factors

Crude oil conversion factors

*Crude Oil**	*To*	*tonnes (metric)*	*kilolitres*	*barrels*	*US gallons*	*tonnes/ year*
From	Multiply by					
Tonnes (metric)		1	1.165	7.33	307.86	–
Kilolitres		0.8581	1	6.2898	264.17	–
Barrels		0.1364	0.159	1	42	–
US gallons		0.00325	0.0038	0.0238	1	–
Barrels/day		–	–	–	–	49.8

*Based on worldwide average gravity.

Natural gas conversion factors

Natural Gas & LNG	*To*	*billion cubic metres NG*	*billion cubic feet NG*	*million tonnes oil equivalent*	*million tonnes LNG*	*trillion British thermal units*	*million barrels oil equivalent*
From	Multiply by						
1 billion cubic metres NG		1	35.3	0.90	0.73	36	6.29
1 billion cubic feet NG		0.028	1	0.026	0.021	1.03	0.18
1 million tonnes oil equivalent		1.111	39.2	1	0.805	40.4	7.33
1 million tonnes LNG		1.38	48.7	1.23	1	52.0	8.68
1 trillion British thermal units		0.028	0.98	0.025	0.02	1	0.17
1 million barrels oil equivalent		0.16	5.61	0.14	0.12	5.8	1

Units of measure

1 metric tonne = 2204.62 lb. = 1.1023 short tons
1 kilolitre = 6.2898 barrels
1 kilolitre = 1 cubic metre
1 kilocalorie (kcal) = 4.187 kJ = 3.968 Btu
1 kilojoule (kJ) = 0.239 kcal = 0.948 Btu
1 British thermal unit (Btu) = 0.252 kcal = 1.055 kJ
1 kilowatt-hour (kWh) = 860 kcal = 3600 kJ = 3412 Btu

Calorific equivalents

One tonne of oil equivalent equals approximately	
Heat units	10 million kilocalories
	42 gigajoules
	40 million Btu
Solid fuels	1.5 tonnes of hard coal
	3 tonnes of lignite
Gaseous fuels	See natural gas and LNG table
Electricity	12 megawatt-hours

Notes

One million tonnes of oil produces about 4,500 gigawatt-hours (= 4.5 terawatt hours) of electricity in a modern power station.

Tonnes = metric tons.

The preferred units used through this volume are as follows:
Primary energy: Exajoules (EJ) or millions of tonnes of oil equivalent (mtoe)
Crude oil: Millions of tonnes (mt)
Natural gas: cubic metres (billions of cubic metres - bcm, trillions of cubic metres-tcm)
Coal: Metric tonnes
Pipeline flows: MMSCMD (million standard cubic meters a day)

References

British Petroleum Conversion Factors, viewed 13 june 2008<http://www.bp.com/conversion factors.jsp>.

Index